I0814693

SURVEYING
THE EARLY REPUBLIC

LIBRARY OF SOUTHERN CIVILIZATION

SURVEYING
THE EARLY REPUBLIC

THE JOURNAL OF

ANDREW ELLICOTT

U.S. BOUNDARY COMMISSIONER
IN THE OLD SOUTHWEST
1796–1800

EDITED BY

ROBERT D. BUSH

LOUISIANA STATE UNIVERSITY PRESS
BATON ROUGE

Published by Louisiana State University Press

Manufactured in the United States of America
First printing

Designer: Barbara Neely Bourgoyne
Typeface: MillerText
Printer and binder: McNaughton & Gunn, Inc.

Library of Congress Cataloging-in-Publication Data
Names: Ellicott, Andrew, 1754–1820, author. | Bush, Robert D, editor.
Title: Surveying the Early Republic : the Journal of Andrew Ellicott, U.S. Boundary Commissioner of the Old Southwest, 1796–1800 / edited by Robert D. Bush.
Other titles: Journal of Andrew Ellicott
Description: Baton Rouge : Louisiana State University Press, [2016] | Series: Library of Southern civilization | Includes bibliographical references and index.
Identifiers: LCCN 2016005176| ISBN 978-0-8071-6342-9 (cloth : alk. paper) | ISBN 978-0-8071-6343-6 (pdf) | ISBN 978-0-8071-6344-3 (epub) | ISBN 978-0-8071-6345-0 (mobi)
Subjects: LCSH: Ellicott, Andrew, 1754–1820—Diaries. | Ellicott, Andrew, 1754–1820—Travel—Southwest, Old. | Surveyors—United States—Diaries. | Southwest, Old—Description and travel—Sources. | Mississippi River Valley—Description and travel—Sources. | Southwest, Old—Historical geography—Sources. | Southern boundary of the United States—History—Sources. | Surveying—United States—History—Sources. | Geographical positions—United States—History—Sources.
Classification: LCC F213 .E4813 2016 | DDC 917.6/02—dc23
LC record available at https://lccn.loc.gov/2016005176

The paper in this book meets the guidelines for permanence and durability of the Committee on Production Guidelines for Book Longevity of the Council on Library Resources. ♾

CONTENTS

ACKNOWLEDGMENTS

The era of United States' expansion into the Southwest after 1790 has attracted, and fascinated, scholars and students for some time. For educated persons in the eighteenth and nineteenth centuries, the "frontier" held a special interest. The "West," or frontier, was more than a direction: it offered adventure, the pursuit of further knowledge about man in his environment, and an unlimited potential for success, or, possibly, failure. Information about the frontier—however far away that might be—was sought with the same degree of intensity and dedication during the eighteenth and nineteenth centuries that our quest for knowledge about space and the universe is pursued today. The opening of the trans-Appalachian West, both north and south of the Ohio River, was an American saga combining elements of audacity, bravery, dedication to a national mission, imagination and foresight for its future potential, individualism, and political ambition, even rebellion, on an international scale. For any of these reasons, and many others, the two decades of history in the Old Southwest from 1783 to 1803 witnessed some of the most important events in American history.

One of the subject areas in this panorama that continues to attract inquiry and study is that of the relationship between Spain and the United States in the Lower Mississippi Valley during these years. Much has been written on Spanish settlement in North America and the Anglo-Spanish contest for global dominance in the eighteenth century. France's departure with her loss of Canada to the British in 1763, along with the British conquest of East and West Florida from Spain, and France's cession of Louisiana to Spain in 1762, left only Britain and Spain, with their respective provinces, in contention for dominance in North America. Spain's subsequent independent successes in the Old Southwest during the American Revolution, with the conquests of

West and East Florida, restored its presence, while leaving Britain holding on to Canada. But after 1783, there was another rival, the young United States. It is this latter story of Spanish-American rivalry that has prompted much research from scholars. We are indebted to several colleagues for their contributions to our understanding of this subject, and to the intellectual pursuit of knowledge in the field of their respective interests in the Mississippi Valley and the Southwest. They inspired us through their love of the subject, and their intellectual diligence. Among them, it is possible here to mention only a few, but they include: Abraham P. Nasatir, John Preston Moore, John Francis Bannon, Charles E. O'Neill, Robin F. A. Fabel, Eric Beerman, John Francis McDermott, Jack D. L. Holmes, Gilbert C. Din, Light T. Cummins, Thomas D. Watson, Arthur Preston Whitaker, and William "Bill" Coker.

SURVEYING
THE EARLY REPUBLIC

EDITOR'S INTRODUCTION

Andrew Ellicott (1754–1820) was born into a Quaker family in Pennsylvania, and, after working in his father's and then his uncle's mills, he turned his attention to science, mechanics, mathematics, surveying, and town or city planning. His work as a surveyor and city planner was such that he came to the attention of President Washington and the Federal Building Commission, established for the creation of the new Territory of Columbia and the City of Washington. It was his mastery of surveying skills and dedication that led President Washington to appoint Ellicott in 1796 as the American commissioner for determining the boundary of the United States with Spain, per the Treaty of San Lorenzo in 1795, in the Old Southwest. During this four-year project (1796–1800), he successfully completed that assignment and filed his "report," the results of which he compiled in 1802 and then published in 1803 from notes entered into his journal.

Ellicott originally published *The Journal* (Philadelphia: Budd and Bartrum, 1803) as a narrative that described his day-to-day activities, and included an appendix of his charts, maps, astronomical, thermal observations, and other scientific data. It was reprinted in 1814, but without some of the maps, charts, and data of the 1803 edition.[1] The 1814 edition, compiled in ten chapters and accompanied by related documents, constitutes the body of this text. It includes, in addition to the narrative of Ellicott's travel, diplomatic correspondence, personal letters, and communications between the participants from both the United States and Spain on site in the Old Southwest. Scholarly interpretations from historians have been added for reference on different subjects as found in the text. Pagination is by sequential numbers provided for the entire text throughout for each page for purposes of reference. Ellicott's original spelling of names and terms (he was a phonetic speller), along with contemporary capitalization, punctuation, paragraph structure, and wording, have been retained. Where necessary, corrective or explanatory notes have been added, either as notes or simply by use of brackets at their point of origin in the text.

We must ask ourselves a logical question at this point: What is the importance of Ellicott's work for our understanding of American history and scientific knowledge at the end of the eighteenth century? As with other contemporary historical memoirs, journals, letters, and reminiscences, Ellicott's writings contribute a body of information and observations that collectively present significant information and interpretation of the day. His work, however, constitutes more than just individual reminiscences; rather, it is a window into the world of the American Enlightenment. Ellicott's work includes important and never before described scientific data. *The Journal* is therefore an excellent overview of various aspects of the United States as it expanded into the southern and western frontiers: its politics, economics, religion, social or community life, intellectual or scientific values and trends. For many reasons, it is worthy both as historical literature and knowledge for the future.

First, *The Journal* provides insight into the daily lives of settlers and the tumultuous history of the Old Southwest during the 1790s. The Old Southwest, land south of the Ohio River and east from the Mississippi River to existing state borders, and south to the border with Spanish territory, was a frontier established as an incorporated territory and designated by Congress as the Southwest Territory in May 1790. The Old Southwest was created from land ceded to the United States by the States of Virginia, North Carolina, and Georgia, and it was a political vacuum into which contending nations, groups, and individuals lived out their ambitions. Throughout his writing, Ellicott meticulously describes communities that he visited, the populations therein, the living conditions, crops and livestock production, Indian relations, and much more. For students of American history, Ellicott's *Journal* is a panorama of what was happening during these expansionist years of the early republic (1796–1800) and, in particular, the tumult on the southern and western frontiers. Included in his narrative are events such as Senator William Blount's "conspiracy," the secret activities of persons known or suspected of being in the service of either Spain (General James Wilkinson) or Britain (Blount), the open uprising in 1797 by the Natchez inhabitants against the Spanish governor Don Manuel Gayoso de Lemos, and the political activities of individual adventurers such as William Augustus Bowles with his Indian allies are all discussed in meticulous detail.

Second, *The Journal* includes copies of correspondence and conversations recorded between the participants, both Spanish and American, about their actions, motives, and contributions—both positive and negative—for the mission objective of the boundary survey and its international repercussions. At key points in the text, Ellicott provided copies of correspondence and reports of meetings between himself and Spanish officials, American officers, Indian chiefs, and settlers whose political views reflected their respective loyalties to the United States, Great Britain, and Spain—or their own selfish motives. Documents quoted include the extensive correspondence between Commissioner Ellicott and Don Manuel Gayoso de Lemos, governor of the Natchez District for Spain. Gayoso was fluent in English, which he learned while a student in Britain, and, in addition, he was married to an American from Philadelphia. The Ellicott/Gayoso correspondence is key to understanding their nations' relationship and respective positions.

Third, Ellicott was a scientist as well as a surveyor, and his work and empirical observations, which he meticulously recorded and disseminated for posterity, not only contributed new information but fostered as well a thirst among his colleagues for more knowledge about the frontier. Ellicott was encouraged in his work by fellow members of the American Philosophical Society—James Madison, David Rittenhouse, Benjamin Franklin, and others—the foremost academic and scientific institution in the country at that time. Ellicott lived and worked during the American Enlightenment, an age of scientific and intellectual excitement that, like its eighteenth-century European counterpart, promoted the advancement of humanity through education and scientific knowledge. The scientist president Thomas Jefferson launched several scientific expeditions during his term of office (1801–9). Upon completion of the boundary survey in 1800, Ellicott returned to work in Philadelphia, where, by the spring of 1804, his scientific reputation was such that Meriwether Lewis studied the principles and techniques of celestial navigation under his direction. Ellicott's scientific observations on the rivers, topography of the area, plant life, Indians, navigation of rivers as well as the Gulf of Mexico, and medical practices were first published in 1803. Much of what he described was complemented by similar observations from Garrett Elliot Pendergrast's, *A Physical and Topographical Sketch,* which was also published in 1803. Taken together or individually, Ellicott and Pendergrast provide fascinating descriptions of scientific inquiry and the advancement of knowledge for everyday life within the world of the American Enlightenment.

Fourth, Ellicott was a Federalist whose general political views reflect the ideology of that political party during the administration of President John Adams (1797–1801). Throughout *The Journal,* Ellicott makes political observations that illustrate the views he shared with other Federalists: he criticizes the Spanish administration for its perceived incompetency; expresses suspicions about French interference with and dominance over Spain in North America; describes the French Revolution as mob rule; portrays the Indians as obstacles to American expansion; praises the American system of liberty; and warns his government repeatedly about the political vacuum in the Old Southwest that he felt had been created by Spain's feebleness, and thereby opening the door to foreign intervention and opportunism.

And, fifth, Ellicott's repeated observations on the political and military weaknesses of the Spanish administration in the Old Southwest provided both contemporaries and the next administration, that of Thomas Jefferson, with a wealth of information. His observations and recommendations focused especially on the importance of future American economic interests in the lands east of the Mississippi River; he described the prospects for American fortifications, trade, and commerce along the rivers and the Gulf of Mexico; and navigational observations, as well as the excellence of the harbors at Mobile and Pensacola.[2] Ellicott's comments and reports on Spain are reinforced by the English political pamphleteer William Cobbett, who, writing under the pseudonym "Porcupine," published several documents by Ellicott to Pickering from the Department of State archives that were widely distributed. Finally, the passage from the British merchant Francis Baily's *Journal,* the appendix to this volume, provides a sharply critical case study of Spanish business practices and political economy in North America in 1797–98. Baily was in Natchez at the same time as Ellicott.

In his preface to the 1803 printing of *The Journal,* Ellicott described his intention and the methodology of his work:

> When *the Journal* was drawn up, it was my intention to have divided the map of the Mississippi river from the mouth of the Ohio, to the Gulf of Mexico, into two parts only; but when I came to lay it down, it appeared better to divide it into three: the map containing the third, or lower part, contains also the island of Orleans, and the greater part of West Florida. The maps belonging to *the*

> *Journal* are all laid down by a scale of 15 miles to an inch, and may easily be annexed to each other by a little attention to the meridians, and parallels of latitude.[3]

A brief discussion includes earlier travel accounts of the Mississippi River, particularly that by Father Louis Hennepin, a Franciscan friar [1640–1701?], which led Ellicott to conclude: "I am inclined to believe the whole to be a fiction . . . that he, and his two men, were captured by the Indians as stated by himself, which circumscribed his excursions, and left a chasm in his discoveries, which he afterwards thought proper to fill up with this relation."[4] In his preface, Ellicott comments also on the recent acquisition of lands on the west bank of the Mississippi with the Louisiana Purchase in 1803, which, at the time of his writing, was the news of the day. As a good Federalist, and a man from the East, he wrote: "On the advantages to be derived to the United States from this cession, there will probably be a great variety of opinions: the security of the navigation of the Mississippi, is certainly an object of the last importance to the inhabitants of our western country, and without which it might be difficult to retain them in the union; but on the other hand, an immediate possession, and sale of the lands west of the Mississippi, might have a tendency to scatter our citizens, already too widely extended to experience all the advantages of society, civilization, the arts, sciences, and good government, and lower the price of our public lands by bringing too great a quantity to market."[5]

Due to the length of *The Journal* (299 pages) and its appendix (151 pages), only the former is included here, with a focus on Ellicott's accounts of travel, encounters with Spanish officials and local population, views on the Indians, political plots, some scientific observations of the area, daily life on the frontier, and the general progress of the boundary survey itself. *The Journal* incorporates a portion of Ellicott's scientific data (temperature variation, astronomy, water quality, soil fertility, crops, navigation on rivers and streams, animal life, maps, and weather conditions) in numerous locations throughout the text.

In 1962, Quadrangle Books published a copy of the 1814 edition of *The Journal* as volume 7 of its American Classics Series, but without a scholarly introduction, any explanatory notes, or bibliography. This volume includes such features, as well as maps, in order to illustrate the process and history involved in the determination of the final boundary between the United States and Spanish possessions remaining in West and East Florida along the line at 31° north latitude. Despite Ellicott's often obvious prejudices, *The Journal* constitutes a useful text, supplemented by the additional documents from contemporaries. The illustrations, especially the maps, extensive notes, and bibliography, provide quick references for students of American history, U.S. diplomatic history, frontier life in the West, and American intellectual and scientific studies of the eighteenth century. *The Journal* provides a window into the world of the ever-expanding American frontier between 1796 and 1800, and the trials and tribulations related to that expansion. Readers are introduced to some of the more fascinating individuals and events in American history: General James Wilkinson, Senator William Blount, the English, although American-born, adventurer William Augustus Bowles, and the 1797 "commotion"—bloodless revolution—on the part of the American inhabitants of Natchez against the Spanish administration. In addition, *The Journal* details such international intrigues as Spanish officials seeking to delay the boundary adjustment per the 1795 Treaty of San Lorenzo because of their fear of a British invasion into Upper Louisiana, New Orleans, and even toward Mexico.

A useful source of documentary information on the events is provided by the three volumes found in Record Group 76 of the National Archives and titled "Records Relating to the Southern Boundary of the United States," comp. Daniel T. Goggin. In this Record Group, official dispatches, reports, and correspondence with enclosures are filed by name of the sender (Ellicott) to recipient (e.g., Secretary of State Pickering), and are listed in chronological order by date of the communication in each of the three volumes. This source is cited in the notes by document title and date, "Southern Boundary," and volume number.

The United States' achievement of independence from Great Britain in the Peace of Paris (September 3, 1783) ended the American Revolution, but it took another year to end hostilities between Great Britain and Spain. For Britain, severing her administrative and colonial ties with the former colonies turned out to be a blessing in that she no longer had to bear the huge administrative costs to maintain the colonies, keep peace between the settlers and the Indians, or negotiate boundaries between the colonies and their neighbors. After all, a boundary dispute between the Colony of Virginia and the French in the western area claimed by both, after years of bloody conflict on the frontier, culminated in the French and Indian War (1754–63). This confrontation, in turn, led to a general global conflict for domination in both Europe and around the globe that lasted for seven years, from 1756 to 1763. With both France and Spain later becoming belligerents against Britain during the American Revolution, peace after 1783–84 simply increased the boundary problems that Britain bequeathed to the United States. Not only did the United States have a boundary issue with Britain regarding Canada and the Great Lakes after 1783, with British troops still occupying forts there, but the "Mississippi Question" with Spain became an increasingly troublesome one. The old boundary line between the British colonies and Spain's occupation of the West Bank of the Mississippi, which Britain had received from France back in 1763, was the center of the Mississippi River channel. In the Peace of Paris (1783) between Britain and the United States, Britain transferred this boundary line designation and all lands east from there to its former colonies.

Spain, however, was not a signatory to this peace treaty. If the Mississippi River boundary line had continued to the Gulf of Mexico, it would have made for a mutually acceptable state of affairs between Spain and the United States. However, the Spanish had won back their former provinces of West and East Florida during the war with Britain (1779–84). In their separate treaty in 1784, Spain claimed the entire channel of the Mississippi River from north of Natchez at the Yazoo River southward to the Gulf. During its years of occupation of West Florida (1763–83), Britain had gradually extended the original boundary from its original designation at 31° north latitude northward to a line near the junction of Yazoo River (32°28′ north latitude) with the Mississippi at a site north of present-day Natchez. From there, West Florida ran due east to the Chattahoochee River, then the boundary ran south along the middle of the river channel to its' junction with the Flint River, and from there straight eastward to the headwaters of the St. Mary's River, and finally down the middle of St Mary's channel to the Atlantic Ocean. Per its separate treaty (1784) with Britain, which did not define boundaries, Spain regained its former Florida properties and claimed that the British boundary in 1784 superseded the former one of 1763 (31° north latitude) that had once separated Spanish East and West Florida from Britain's American colonies.[6] After 1784, therefore, the United States was blocked from access to the Gulf of Mexico since the Mississippi River north of Natchez to the Yazoo, and then south, along with the port of New Orleans, was now firmly in the hands of Spain. In addition, by 1795, Spain had established forts and small settlements up the Mississippi River as far north as St. Louis.[7] The Gulf of Mexico was a Spanish lake. To make matters worse for the Americans, in 1784 to protect its interests, the Spanish government officially closed its portion of the Mississippi and the port of New Orleans to all commercial traffic except that of Spanish citizens.

The United States after 1783 faced similar boundary problems also within its own territories, the rival land claims between the various states with regard to their respective boundaries of the western lands. Virginia claimed land as far north as what is today Michigan based upon its former royal charters. All of the states claimed land westward to the Mississippi River, beyond which, everyone recognized, the land belonged to Spain. And, compounding the problem, the American government under the Articles of Confederation was powerless to intervene in disputes between states. Virginia and Maryland, for example, almost went to war over rival navigation claims on the Potomac River. The decision, beginning in 1784, on the part of the states to relinquish their conflicting land claims to the Congress was one solution. With its efforts to resolve the issues of boundaries, the Congress passed land ordinances in 1784, 1785, and 1787 (the Northwest Ordinance). All of these concessions by the states made strides toward resolution. But the sale of the western lands to developers and individuals depended upon accurate surveys, despite the plan of the 1787 Northwest Ordinance for division into townships and sections of land allocation. Land relinquished south of the Ohio River and west to the Mississippi River from the state boundaries of Virginia, North Carolina, and Georgia and along their southern boundaries with Spanish properties was designated by Congress as the Southwest Territory in May, 1790. Along with the other problems confronting the early American republic under the Articles of Confederation, the issue of boundaries in the West and South had to await resolution by a new federal government under the Constitution. None of these boundaries issues was more important to the future of the United States, because of the increased American immigration into the trans-Appalachian frontier, than the resolution of the navigation on the Mississippi River, and the boundary with Spanish West and East Florida. Since no boundaries had ever been surveyed, no one had any realistic idea of the extent of these western lands, nor were there any reliable maps.[8]

Surveys and surveyors were therefore desperately needed, and one person who met this need was Andrew Ellicott. He was born January 24, 1754, in Bucks County, Pennsylvania, and died on August 28, 1820, at West Point, New York, where he was professor of mathematics at the West Point Military Academy. Ellicott was the firstborn of nine children and was raised and educated in the local Quaker community. He worked in his father's mill and, later, in one operated by his father and two uncles; but he was also skilled in mechanics, mathematics, and science. In 1775, he married Sarah Brown (1756–1827) of Newtown, Pennsylvania, with whom he had nine children. When the American Revolution broke out, despite his Quaker background, Ellicott enlisted in the Maryland militia and achieved the rank of major. When the war was over, he returned to Pennsylvania, where he worked on local surveys and became active in scientific circles. He and his family moved to Baltimore in 1785, when he was appointed as a teacher in the Baltimore Academy, and from there, in 1786, he was assigned the task to survey the western border for the State of Pennsylvania by the state legislature. In 1789, he again moved, this time to Philadelphia. As a Federalist in the capital city, he took the job to survey and record lands near Lake Erie in order to determine the western borders between Pennsylvania and New York. Ellicott's survey work, including the first recorded description of Niagara Falls, established his reputation for accuracy. In 1791–92, he surveyed the boundaries of the Territory of Columbia, later the District of Columbia. In the process, Ellicott continued to reinforce his reputation for precision and accuracy in city planning through markings of streets and squares for the future City of Washington. He soon, however, became enmeshed in a dispute with Pierre Charles L'Enfant, who initially had been hired to prepare the plans for the city. In 1792, President Washington dismissed L'Enfant because of his constantly changing plans, frequent revisions of schedules, and failure to produce a plat for the city that could be used by potential investors and thereby facilitate the city's construction.[9] Washington entrusted the task to Ellicott in March 1792. Ellicott's survey plan was published and adopted by the Building Commission later that year; it became the outline survey for the City of Washington. His work in the Territory of Columbia finished, he returned to Pennsylvania to work on surveys there. But, thanks to events in Europe, a new need for his services was now on the horizon.

The outbreak of war in Europe in 1793, which stemmed from the French Revolution of 1789, created European upheavals and international tensions. Spain was among the powers involved in the war, but its military defeats by the French (1794–95) prompted it to abandon the anti-French coalition, which included Britain. Among Spain's urgent needs was to realign its international interests and divest itself of unprofitable colonies; quite simply, Spain was on the verge of bankruptcy after years of warfare. One plan included consideration of concessions to the neutral (1793) United States, which the Spanish Supreme Council of State discussed in July 1794; a second idea discussed was to seek peace by a separate treaty with France, which Spain did in the Treaty of Basel in July 1795. By the latter course of action, Spain became, once again, an enemy of Britain because France was still at war with Britain. Knowledge of Anglo-American negotiations and a resulting commercial treaty in November 1794 (Jay's Treaty) confirmed the wisdom of a rethinking of Spain's relationship with the United States. Any prospect of a future Anglo-American military alliance, despite assurances of neutrality in 1793 from the United States, would threaten Spain's possessions in North America, Mexico, and the Gulf of Mexico. Manuel de Godoy (1767–1851), prime minister to King Carlos IV of Spain, anticipated the dangers and warned the Spanish Council of State on July 7, 1794, of such a potential eventuality. He told them that Spain must be prepared and seek an alliance with the United States that "could aid us most and, if it were hostile to us, could do us the greatest harm."[10] Such a possibility of an Anglo-American military alliance in the future was to be prevented at all costs. On October 27, 1795, only four months after the Treaty of Basel with France, Spain and the United States signed the Treaty of San Lorenzo (Pinckney's Treaty in the United States), which professed peace and friendly relations, recognized the need to define their respective boundaries, opened navigation on the Mississippi to the Americans, and for other purposes.[11] These provisions, Spain believed, would neutralize the United States, in that the treaty responded to the American's principal concern, free navigation of the Mississippi River and access to the port of New Orleans. On March 7, 1796, the treaty was ratified by the United States Senate. Article II stated in part: "The Southern boundary of the United States which divides their territory from the Spanish Colonies of East and West Florida, shall be designated by a line beginning on the River Mississippi at the Northernmost part of the thirty first degree of latitude North of the Equator, which from thence shall be drawn due East to the middle of the River Apalachicola or Catahoucha [Chattahoochee], then along the middle thereof to its' junction with the Flint, then straight to the head of the St. Mary' River, and thence town the middle thereof to the Atlantic Ocean." Article III stated "in order to carry out the preceding article into effect one commissioner and one surveyor shall be appointed by each of the contracting parties."

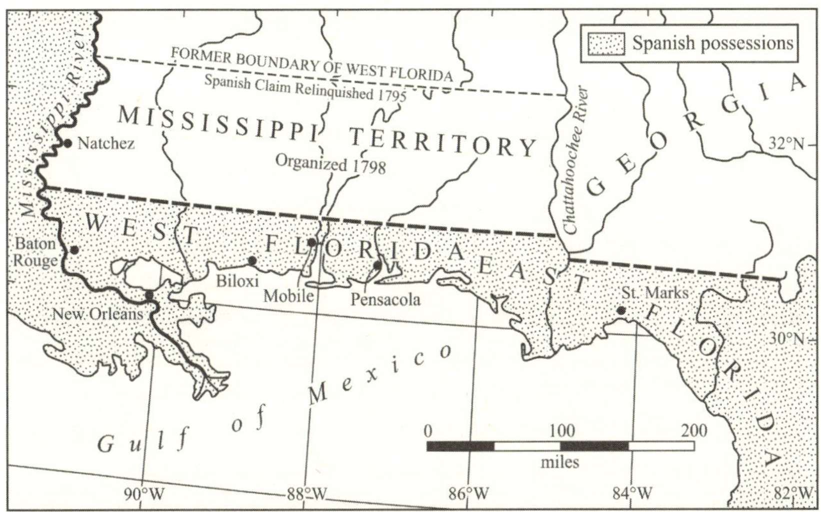

Final map on the boundary line between the United States and Spanish possessions at 31° north latitude for West and East Florida to the Atlantic Ocean. Detail from 1803 portion of map in Perry-Castañeda Library Map Collection, University of Texas at Austin Libraries. Courtesy Wikimedia Commons, the free media repository.

The United States therefore needed to meet its treaty requirement per Article III, and President Washington had just the right person in mind to act as the American commissioner. On May 21, 1796, he appointed Andrew Ellicott, a fellow Federalist, as commissioner and Thomas Freeman as surveyor to represent the United States for the determination of the agreed-to boundary line of 31° north latitude between the United States and Spain. Ellicott's appointment was ratified by the Senate on May 24, 1796, along with an appropriation of thirty thousand dollars from Congress with which to cover expenses. The treaty became effective on August 2, and Ellicott departed Philadelphia on September 16 overland to Pittsburgh, and then via the Ohio and Mississippi Rivers to Natchez, the Spanish district seat of government, where he arrived on February 24, 1797.[12] Prior to his arrival in Natchez, Ellicott gathered information from a number of sources as to the situation awaiting him. His initial encounters with Spanish officers stationed at Chickasaw Bluffs (near present-day Memphis) and their other forts along his route were a prelude to his future troubles in that the local Spanish officers feigned ignorance of the treaty and attempted to delay his travel. Such impressions confirmed his suspicion that Spain did not intend to honor its obligations.[13]

In fact, however, the Spanish court and Prime Minister Godoy had ordered its officers in the Province of Louisiana to comply with the terms of the treaty in August 1796. But Spanish officials such as Governor-General Baron de Carondelet in New Orleans, Governor Manuel Gayoso de Lemos in the District of Natchez, and Marquis de y Riujo, Spanish minister to the United States in Philadelphia, all believed the treaty to be a mistake. Being on the local scene, these Spanish officers were in a position to see firsthand the pro-British attitude of the Federalist government and its press following the announcement of Jay's Treaty. Simultaneously Carondelet and Gayoso dealt daily with Americans on the southwestern frontier, some of whom were opposed to the East Coast Federalists. Orders from Spain were, however, orders!

Ellicott's early relations with Governor Gayoso in Natchez were friendly, and although he continued to be suspicious of Spanish intentions, he nevertheless took actions to ensure that there would be no misunderstanding of his determination to complete his mission.[14] But world events altered Spanish policy; in October 1796, new orders were drafted in Spain and arrived in the Province of Louisiana in March 1797. The new orders from Prime Minister Manuel de Godoy instructed Spanish officers in Louisiana to delay implementation of the treaty because Godoy and his French ally both suspected the motives in Jay's Treaty between the United States and Britain. Was this treaty not a prelude to a military alliance? Governor Don Manuel Gayoso de Lemos initiated Spanish delays in implementing the treaty, utilizing issues such as the need to repair defenses against a possible British invasion, or the disputed property rights for settlers holding Spanish land grants in the Natchez District and elsewhere. On March 29, 1797, Gayoso issued a proclamation to the inhabitants of Natchez. Included as a document in Ellicott's *Journal,* the proclamation read, in part: "I find it my duty to explain, that a negociation is now carrying on to secure the right of the said real property. As that right cannot be secured but by an additional article to the late Treaty, and until that article is officially communicated to me I am bound to keep possession of the Country, and continue to its inhabitants the same indulgence & the same anxious protection as until now." Gayoso cleverly worded his decision to delay the implementation of the treaty because of concerns over the property rights of the inhabitants who held Spanish land grants. Any new negotiations in the respective capitals would take months, if not years, to complete. Spain, an ally of France, was still at war with Britain, which raised the possibility of a future Anglo-American military alliance. It was well known in Spain and France that the Federalist government and its press in the United States was decidedly pro-British. Any delay in implementing the treaty with the United States would therefore provide Spain with an opportunity to reassess the situation and, if necessary, make further defensive preparations. In Spain, however, Prime Minister Godoy believed that their treaty with the United States, as it currently stood, would keep the Americans in line because of the latter's desire for navigation rights on the Mississippi. In the meantime, Spanish defense works were strengthened against the eventuality of a possible invasion from the north.

At this critical moment, 1797, Ellicott was deeply involved with political events in Natchez. He was convinced from local intelligence and his own observations gathered on his travel down the Ohio and Mississippi Rivers that Spanish officers did not intend to comply with the treaty. Gayoso's proclamation, Ellicott believed, was merely smoke and mirrors to hide Spain's real intentions: to avoid transferring any territory to the United States and completing the boundary survey. As he confessed to Secretary of State Timothy Pickering, he had been privy, after only a short time in Natchez, to a confidential letter (June 16, 1796) in which Governor Gayoso "assures his friend [Daniel Clarke Sr.] that the court of Madrid had no design of carrying the Treaty into effect further than what related to the navigation of the Mississippi." The events narrated in *The Journal* from February 1797 until the final evacuation of the Spanish posts, including Natchez, in March 1798, therefore led Ellicott to take the offensive in serving his country's interests as he saw them. But suspicion of Spanish intransigence was not the only factor; at the same time, into the political vacuum in the Old Southwest created by Spain's weaknesses, entered pro-British agents, individual plots and conspiracies by Americans, Indian maneuvering, and local unrest in Natchez. Arthur Preston Whitaker has described Ellicott's position at this time:

> In the course of this year almost every thread of frontier history was gathered up at the tiny post on the Lower Mississippi. Spanish conspirators of Tennessee and Kentucky, promoters of land speculation at Muscle Shoals and in the Yazoo country, officials of the rival governments, Indians and Indian agents, and ringleaders of the Blount conspiracy—all met in the little town that lay between the river and the worthless Spanish fort on a hill nearby. Though the population of the town and the surrounding district was not large, the behavior of the people

> was of vital importance; and they were so heterogeneous a mass—Spaniards, Frenchmen, Britons, and Americans from many states—that public opinion was unpredictable from one week to the next. If, as Daniel Clark [Sr.] wrote somewhat later, these people were always "restless and turbulent," the events of 1797 gave them plenty of the action that they found so congenial.[15]

For three and one-half years, Ellicott would be engaged in the boundary project, although the actual survey work on the boundary line would not begin until more than a year after his arrival (April 9, 1798) because of political events that caused the Spanish to delay their portion of the survey work. In the meantime, Ellicott provided his government with a wealth of information on the Old Southwest, the area west of the Appalachians south of the Ohio River; on events in the states of Kentucky, which had entered the Union in 1792, and Tennessee, which had entered in 1796; and on future boundary issues.

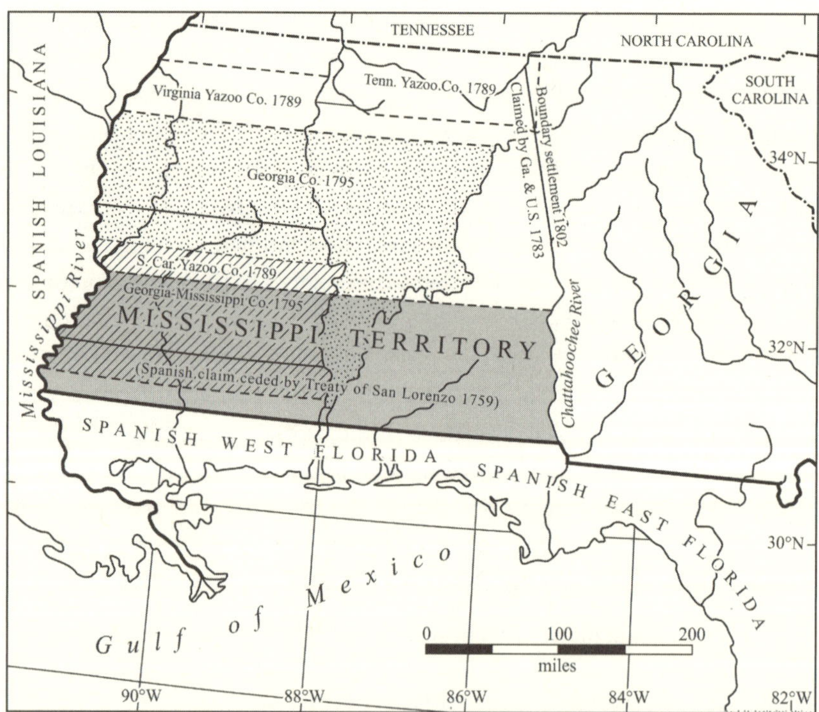

"Georgia's Western Lands 1783–1803." This map identifies the boundaries of the Mississippi Territory, created by Congress in 1798, and the features of other boundary claims yet to be resolved by the states. Map from *The American Heritage Pictorial Atlas of United States History* (New York: American Heritage Publishing, 1966), 123.

Ellicott's original instructions included more, however, than simply running a boundary line: "In addition to his formal instructions, he was verbally advised [by President Washington] to watch General Wilkinson, whose intrigues had long since given rise to damaging reports. Conditions at Natchez might cause him [Ellicott] to assume indefinite political functions."[16] He assumed such political functions, which in addition to his specific title as commissioner of the United States for the determination of the boundary between Spain and the United States, prompted him to act on his own as "minister without portfolio" in the ensuing diplomatic relations with Spanish officers and Indian chiefs on site. It was a duty that he relished; there is no doubt, however, that he exceeded his authority and instructions. And the delays were expensive; his original congressional appropriation of thirty thousand dollars was quickly consumed because of delays and additional personnel expenses. A supplemental appropriation of twelve thousand dollars was later added to his authorization.

Ellicott's *Journal,* compiled in 1802 and first published a year later, described local people, fauna and flora, Indian relations, everyday life on the frontier, and, from the beginning of his arrival, included the complex and often vexing relations with the representatives of Spain.[17] As noted earlier, Ellicott obtained a copy of a 1796 letter from Governor Gayoso to Daniel Clark Sr. From that letter, which he quoted often, Ellicott pointed out Spain's clear intention not to complete the terms of the treaty of 1795.[18]

The former British boundary had been from the mouth of the Yazoo River (32°28′ north latitude) and the Mississippi River, then east to the Chattahoochee River, south along its channel to its junction with the Flint River, and from there straight eastward along the Saint Marys River to the Atlantic Ocean. This boundary was claimed by Spain at its peace talks with Britain in 1784. The boundary line, per the 1795 treaty, between the United States with Spain, for West Florida would be returned, however, to that of 1763, or 31° north latitude. After the evacuation of Natchez by the Spanish on March 30, the United States therefore created the area in between as the new Mississippi Territory on April 7, 1798. The seat of the new American government for the Mississippi Territory replaced the former Spanish one in the district capital at Natchez after the latter's departure.

The Journal, along with the excerpts from the work by Baily included as this volume's appendix, therefore offers detailed accounts of conflicting alliances, everyday life among the inhabitants, special-interest groups discussing plots and talk of secession, Indian affairs, medical issues within the region, and additional scientific observations on maritime conditions as well as data on fauna and flora. It chronicles events and circumstances during the presidency of John Adams in the American Old Southwest from 1796 to 1800. Ellicott's situation, which required him to understand and maneuver within the complex political and social fabric of Spanish institutions, called for a political skill that went well beyond his diplomatic and scientific mission. His role during this time has been summed up as follows:

> Although not a Houston or a Fremont, he played a similar part in American Expansion. Finding himself upon disputed territory which his government greatly desired; surrounded by an unsettled population, most of whose elements were eager for American control; facing officials of an expiring regime determined to make a last despairing effort to hold the coveted territory, he furthered every effort to stimulate a revolt [June 1797] against the Spaniards. The resulting insurrection was a bloodless one, but it was none the less effective. Thus the Natchez district served as a prototype for West Florida, as that region in turn did for Texas and California.[19]

Chapters 1 and 2 of Ellicott's *Journal* provide candid observations of the landscape, conditions on the frontier at Pittsburgh and the Ohio River to the Mississippi and then on to Natchez; stops along the way down the Mississippi included his initial encounters with Spanish officials at New Madrid, Chickasaw Bluffs, and Walnut Hills. These early encounters helped to set the tone and his negative opinion of Spanish officials, who not only professed ignorance of the Treaty of San Lorenzo but deliberately tried to retard his movements into their domain. Although the weather and ice on the rivers delayed Ellicott's progress too, the actions of the Spanish fort commanders contributed to the delay of his arrival in Natchez until February 24, 1797. Given a warm reception in Natchez by Spanish governor Don Manuel Gayoso de Lemos (1747–1799), who served in that capacity from 1787 to 1797. Ellicott was determined to establish his presence. He proceeded to set up his camp opposite the Spanish fort and with his escorts publicly hoisted the American flag, despite the fact that the Spanish had not yet evacuated the district.[20] This action was an affront not only to the Spanish but also to the Indians, who looked to Spain as their protector in several peace treaties. The tribes were alarmed by the arrival of the Americans, whom they regarded as a threat to their lands. The Indians had agreed to permit the construction of

Spanish forts, such as that at Fort Nogales, on the condition that they would be used to protect both the Spanish and their Indian allies from American intrusions. Spain abruptly abandoned the Indians in the 1795 treaty, which left them on their own and at the mercy of the Americans. The Old Southwest frontier, after 1795, was up for grabs because of weaknesses in Spain's military and administration, and this soon led to chaos and confusion. As Whitaker concludes: "There can be little doubt that the conduct of the commissioner appointed to represent the United States in the transfer and to survey the international boundary, contributed handsomely to the chaos at Natchez."[21]

Chapters 3, 4, and 5 of *The Journal* chronicle Ellicott's communications and confrontations in Natchez, including a rebellion by the inhabitants against the Spanish in June 1797 in which Ellicott played a role. The decision by Governor Gayoso to accept a list of conditions from them, and a similar acceptance by Governor-General Carondelet in New Orleans, brought actual Spanish administration in the Natchez District to an end. Included in these chapters are copies of the extensive correspondence with the Spanish, dispatches to Secretary of State Pickering, and communications between Ellicott and local citizens. In September 1798, the government in Madrid ordered the evacuation of all remaining forts north of the boundary line as required by the treaty, and the matter was concluded in 1800.[22] It was only after the decision by the Spanish to vacate Natchez and Walnut Hills that the actual boundary survey finally began on April 9, 1798.

Chapters 6 through 10 provide detailed accounts, observations and markings of the boundary survey, but without any final markers in some areas because of hostile Indians in the Creek and Seminole lands of East Florida.[23] Among the events chronicled in *The Journal* was Ellicott's meeting with the English agent and self-appointed Indian organizer/adventurer William Augustus Bowles, whose mission as a covert operative of Britain was to once again stir up Indian opposition to the Spanish.[24] This encounter is described in addition to Ellicott's reports on the activities of General James Wilkinson, and the local repercussions of the conspiracy to detach a portion of the West from the Union by Senator William Blount.[25] Between 1798 and 1799, once Ellicott became finally engaged in running the boundary line, he chronicled his daily actions and activities in *The Journal*. The process of the boundary survey provided Ellicott with the opportunity to engage in his scientific observations and notes on the land, rivers, potential sites for forts, and navigation updates for maritime use. His observations and comments on the Gulf Coast,

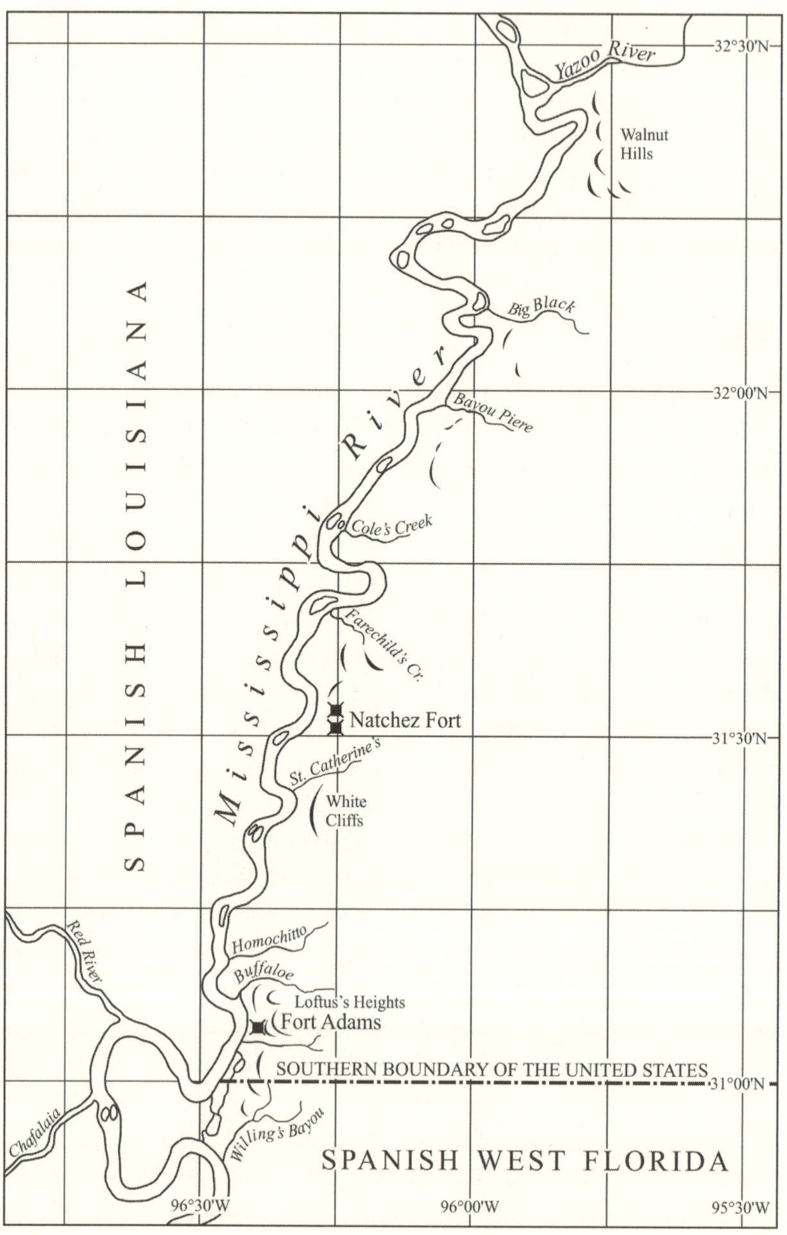

Ellicott's map of the area from 32°28′ north latitude south to 31° north latitude along the Mississippi River and lands eastward, which would be designated by Congress as the new Mississippi Territory in April 1798, following their evacuation by the Spanish on March 31, 1798. Important landmarks on this map below are referenced throughout *The Journal.*

especially on the harbors at Mobile and Pensacola, were important to the later efforts by President Thomas Jefferson and Secretary of State James Madison to acquire "lands to the east" of the Mississippi River for American use. Ellicott finally traveled to St. Augustine, made some observations and notes, and from there returned to Philadelphia, where he arrived on May 18, 1800.

His boundary survey work completed, Ellicott filed his final report with the administration in 1803. Then, as a private contractor, he did survey work for the States of Pennsylvania, Georgia, and North Carolina. After several years of survey and mapping work for the states, he was appointed as professor of mathematics in 1813 at West Point Military Academy, where he remained, with one interruption to assist in the survey work on the U.S. Canadian boundary, until his death on August 28, 1820.

NOTES

1. The full title cited: *The Journal of Andrew Ellicott Late Commissioner on Behalf of the United States during Part of the Year 1796, the Years 1797, 1798, 1799, and Part of the Year 1800: For Determining the Boundary between the United States and the Possessions of His Catholic Majesty in America Containing Occasional Remarks on the Soil, Rivers, Natural Productions, and Diseases of the Different Countries on the Ohio, Mississippi, and Gulf of Mexico, with Six Maps Comprehending the Ohio, the Mississippi from the mouth of the Ohio to the Gulf of Mexico, the whole of West Florida, and part of East Florida. to Which Is Added an Appendix, Containing all the Astronomical observations made use of for determining the boundary, with many others, made in different parts of the country for settling the geographical positions of some important points, with maps of the boundary on a large scale; likewise a great number of Thermometrical Observations made at different times and places* (Philadelphia: printed by William Fry, 1814). Hereafter cited as *The Journal.*

2. After news of the Louisiana Purchase in 1803 was confirmed, Ellicott, who had written on the importance of Mobile and Pensacola as future ports for the United States, concluded: "By the cession of Louisiana to the United States, we gain little on the Gulf of Mexico, and are but little benefitted as a maritime people. The important and safe harbours, in both the Floridas, still remain in the possession of His Catholic Majesty" (*The Journal,* vi). A similar view was echoed in the Federalist press. The *Boston Columbian Centennial* (July 13, 1803) carried an editorial that stated: "We are to give money of which we have too little of for land of which we already have too much." See Jerry W. Knudson, "Newspaper Reactions to the Louisiana Purchase: 'This New Immense Unbounded World,'" *Missouri Historical Review* 63 (January 1969): 182–213.

3. *The Journal,* v.

4. Ibid.

5. Ibid.

6. For an excellent overview of Spanish North America, see David J. Weber, *The Spanish Frontier in North America* (New Haven, CT: Yale University Press, 2009). On the frontier and borderlands issues, see the following recent scholarly studies: Robert V. Hayes, *The Mississippi Territory and the Southwest Frontier, 1795–1817* (Lexington: University Press of Kentucky, 2010); David Narrett, *Adventurism and Empire: The Struggle for Mastery in the Louisiana-Florida Borderlands, 1762–1803* (Chapel Hill: University of North Carolina Press, 2015); and F. Todd Smith, *Louisiana and the Gulf South Frontier, 1500–1821* (Baton Rouge: Louisiana State University Press, 2014).

7. See the classic studies by Isaac Joslin Cox, *The West Florida Controversy, 1798–1813: A Study in American Diplomacy* (Baltimore: John Hopkins University Press, 1918); and Arthur Preston Whitaker, *The Mississippi Question, A Study in Trade, Politics and Diplomacy* (Gloucester, MA: Peter Smith, 1934).

8. See maps 1, 2, and 3. See also the classic study by Samuel Flagg Bemis, *Pinckney's Treaty: America's Advantage from Europe's Distress* (New Haven, CT: Yale University Press, 1965).

9. For details on this matter, see Peter Charles L'Enfant (1791), "Plan of the city intended for the permanent seat of the government of t[he] United States projected agreeable to the direction of the President of the United States, in pursuance of an act of Congress passed the sixteenth day of July, MDCCXC, establishing the permanent seat on the bank of the Potomac" http://memory.loc.gov; Andrew Ellicott (1792), "Plan of the city of Washington in the Territory of Columbia: ceded by the states of Virginia and Maryland to the United States . . . ; engraved by Thakara & Vallance," digitalgallery.nypl.org/nypldigital/dgkeysearchdetail.cfm?strucID-253036; and Elizabeth S. Kite, *L'Enfant and Washington, 1791–1792* (Baltimore: Johns Hopkins University Press, 1929).

10. Speech of Manuel de Godoy, Prince of Peace, to the Spanish Council of State, July 7, 1794, qtd. in Arthur P. Whitaker, "Louisiana in the Treaty Of Basel," *Journal of Modern History* 8, no. 1 (1934): 23. See also Whitaker, *The Mississippi Question,* chaps. 7 and 8. Prime Minister Godoy, in his *Memoirs,* described Spanish action (1795) with the United States in the Treaty of San Lorenzo as being dictated by Britain's global interests, and Spanish fear of an Anglo-American military alliance, following the commercial provisions in Jay's Treaty (1794), for the conquest of its properties in the Antilles and Mexico. See Don Manuel de Godoy, *Memoirs of the Prince of the Peace,* trans. J. B. D'Esmenard, 2 vols. (London: Richard Bentley, 1836), 1:455–60.

11. See Jack D. L. Holmes, "Some Economic Problems of Spanish Louisiana," in *Readings in Louisiana History* (New Orleans: Louisiana Historical Association, 1978), 55–60; and Arthur P. Whitaker, "The Retrocession of Louisiana in Spanish Policy," *American Historical Review* 39 (April 1934): 454–76. On the issue of the slave rebellion, see Laurent Dubois and John D. Garrigus, *Slave Rebellion in the Caribbean, 1789–1804: A Brief History with Documents* (Boston: Bedford/St. Martins, 2006); and David P. Geggus, ed., *The Impact of the Haitian Revolution in the Atlantic World* (Columbia: University of South Carolina Press, 2001).

12. Ellicott's appointment is found in *Senate Executive Journal,* 1:210–11, qtd. in Clarence Edwin Carter, comp. and ed., *The Territorial Papers of the United States,* vol. 5, *The Territory of Mississippi 1798–1817* (Washington, DC: U.S. Government Printing Office, 1949), 3. An excellent summary of Ellicott's activities upon his arrival in Natchez at this time from the viewpoint of Spanish Governor Gayoso is provided by Jack D. L. Holmes, *Gayoso: The Life of*

a Spanish Governor in the Mississippi Territory, 1789–1799 (Gloucester, MA: Peter Smith, 1968), 176–78, 183–87, 231, 233–36, 269–71. Hereafter cited as *Gayoso.*

13. Cox, *The West Florida Controversy,* 33.

14. For an excellent summary of Ellicott's political views of his Spanish counterparts, perhaps Arthur Preston Whitaker has summed it up best: "He never believed, either then [1797] or subsequently, that the Spaniards had the slightest intention of executing the treaty in good faith." Whitaker, *The Mississippi Question,* 59. In a letter from Governor Gayoso to Daniel Clark Sr. dated June 16, 1796, Gayoso was quoted by Ellicott as saying: "And, [he] concludes by declaring in the most unequivocal manner, that it was neither the design of his court, nor the intention of his Catholic Majesty's officers to give efficacy to the Treaty." (Ellicott to Secretary of State Pickering, January 10, 1799, in "Records Relating to the Southern Boundary of the United States," comp. Daniel T. Goggin, 3 vols., Record Group 76, vol. 3; hereafter cited as "Southern Boundary").

15. Whitaker, *The Mississippi Question,* 58. In a letter to Secretary of State Pickering, Ellicott placed the number of inhabitants in Natchez at about 4,000 divided into three categories. He wrote: "The inhabitants, who are much more numerous than I expected, may be divided into three classes. First, those who are either Americans, or those warmly attached to the interest of the U.S. Under this class may be reckoned seven-eighths of the inhabitants. Secondly, old British subjects, who would prefer the government of Great Britain to that of any other nation, but at the same time are decidedly in favor of that of the U.S. when contrasted with that of Spain. Thirdly, about eight persons including some officers, who are attached to the Spanish interest." (Ellicott to Secretary of State, April 14, 1797, in "Southern Boundary," vol. 1).

16. Cox, *The West Florida Controversy,* 33.

17. According to Whitaker, "Neither [Governor] Gayoso's courtesy nor the actual dismantling of the fort [at Natchez] increased his [Ellicott's] trust in Spain. He never believed, either then nor subsequently, that the Spaniards had the slightest intention of executing the treaty in good faith. He was convinced that French influence, if nothing else, would embroil the United States and Spain, for France and Spain after 1795 were allies and war between France and the United States seemed imminent." Whitaker, *The Mississippi Question,* 59.

18. In a confidential letter to Secretary of State Pickering (January 10, 1799), Ellicott reiterated his distrust of the Spanish and their constant delaying tactics regarding completion of the boundary. Ellicott quoted Governor Gayoso's letter of June 16, 1796, to Daniel Clark Sr., which read, in part: "The governor assures his friend that the Court of Madrid had no design of carrying the treaty into effect further than what related to the navigation of the Mississippi. . . . That explanations and adjustments would be required till the Treaty itself would become a dead letter" (in "Southern Boundary," vol. 3).

19. Cox, *The West Florida Controversy,* 41.

20. Ellicott's action by placing the American flag at his encampment was a deliberate act intended to demonstrate his presence. According to his biographer, Jack D. L. Holmes, "Gayoso had no objection to the *principle* of flying the flag, but regarded the timing as poor" (Francis Baily, *Journal of a Tour in Unsettled Parts of North America in 1796 & 1797,* ed. Jack D. L. Holmes (Carbondale: Southern Illinois University Press, 1969), 292.

21. Whitaker, *The Mississippi Question,* 58.

22. Spain's vacillation regarding implementation of their requirements in the Treaty of San Lorenzo is best seen through the eyes and actions of Governor Gayoso in Natchez (see Holmes,

Gayoso, 174–99). In *The Mississippi Question,* Whitaker gives a comprehensive overview of the conflicting orders regarding implementing the treaty from Prime Minister Manuel de Godoy in Spain, and of the position on transfer of the area to the Americans taken by Governor-General Baron de Carondelet in New Orleans, who opposed it (59–67). See also Cox, *The West Florida Controversy,* 32–55; and Bemis, *Pinckney's Treaty.* Spain's position in 1797–98 was dictated by Franco-American relations regarding the latter's ties with Britain per Jay's Treaty. France considered it a breach of American neutrality and began a systematic seizure of American ships in 1797. In retaliation, in 1798, the Federalist-dominated Congress issued letters of marque to American vessels authorizing them to seize French ships, and declared the Franco-American Treaty of 1778 "null and void." This seizure of American vessels by French privateers and vice versa by the Americans resulted in the so-called Quasi-War, which lasted from 1797 to 1800. For Spain's position, see William Cobbett, *Porcupine's Works* ... (London: Printed for Cobbett and Morgan, 1901), vol. 6, July 3, 1797. On the war, see Alexander DeConde, *The Quasi-War—Politics and Diplomacy of the Undeclared War with France, 1797-1801* (New York: Scribner's, 1966).

23. For a look at the Indian nations involved, see Jack D. L. Holmes, "The Choctaws in 1795," *Alabama Historical Quarterly* 30, no. 1 (Spring 1968): 533–49. More lengthy studies of the Indian nations in the Old Southwest and their transition into the United States include Robbie Ethridge, *Creek Country: The Creek Indians and Their World* (Chapel Hill: University of North Carolina Press, 2003); Greg O'Brien, *Choctaws in a Revolutionary Age, 1750–1830,* Indians of the Southeast (Lincoln: University of Nebraska Press, 2005); Claudo Saunt, *A New Order of Things: Property, Power, and the Transformation of the Creek Indians, 1735–1816,* Studies in North American Indian History (Cambridge: Cambridge University Press, 1999); and the excellent study by Charles A. Weeks, *Paths to a Middle Ground: The Diplomacy of Natchez, Boukfouka, Nogales, and San Fernando de las Barrancas, 1791-1795* (Tuscaloosa: University of Alabama Press, 2005). The Choctaws were favorably described by Ellicott; however, he refers on more than one occasion to all Indians as "savages."

24. See the scholarly studies on Bowles provided by Lawrence Kinnaird, "The Significance of William Augustus Bowles' Seizure of Panton's Apalachee Store in 1792," *Florida Historical Quarterly* 9 (1931): 156–92; Elisha P. Douglas, "The Adventurer Bowles," *William and Mary Quarterly,* 2nd ser., 6 (1949): 3–23; Gilbert C. Din, *War on the Gulf Coast: The Spanish Fight against William Augustus Bowles* (Gainesville: University Press of Florida, 2012); and Narrett, *Adventurism and Empire.* Also relevant to an understanding of Bowles is the study by William S. Coker, *Indian Traders of the Southeastern Spanish Borderlands: Panton, Leslie and Company and John Forbes and Company, 1783-1847* (Gainesville: University Press of Florida, 1966). Finally, there is Bowles himself: William Augustus Bowles, *Authentic Memoirs of William Augustus Bowles, Esquire, Ambassador from the United Nations of Creeks and Cherokees to the Court of London* (London: R. Foulder, 1791).

25. Isaac Joslin Cox, ed., "Documents on the Blount Conspiracy, 1795–1797," *American Historical Review* 10 (1905): 574–606; and, on General James Wilkinson, see Andro Linklater. *An Artist in Treason: The Extraordinary Double Life of General James Wilkinson* (New York: Walker, 2009); Arthur P. Whitaker, "James Wilkinson's First Descent to New Orleans in 1787," *Hispanic American Historical Review* 8, no. 1 (February 1928): 82–97; and James Wilkinson, *Memoirs of My Own Times,* 3 vols. (Philadelphia: Abraham Small, 1816).

ANDREW ELLICOTT CHRONOLOGY

1754, January 24	Born in Bucks County, Pennsylvania
1775	Married Sarah Brown
1777–83	Militia officer, rank of major, from Maryland
1784	Appointed as surveyor for the Mason-Dixon Line
1786	Appointed to survey western border of Pennsylvania by state legislature
1789	Appointed to survey southwestern boundary of New York
1791	Appointed as surveyor for the federal Territory of Columbia (later [1801] District of Columbia), Washington City
1792	Appointed by President Washington to complete L'Enfant's work
1792	Published first "Plan of the City of Washington in the Territory of Columbia" (later District of Columbia)
1794–96	Appointed by State of Pennsylvania to plan the fort road from Reading to Fort Erie
1796, May 21	Appointed by President Washington as the U.S. commissioner for the survey of the U.S. and Spanishborder at 31° north latitude per the 1795 Treaty of San Lorenzo
1797, February 24	Arrived in Natchez, Spanish district seat of government, and set up camp
1797, June	Upheaval in Natchez against Spanish administration
1798, March 31	Spanish evacuate Natchez per the Treaty of San Lorenzo
1798, April 9	Joint Spanish-American survey work begins
1800, May 18	Returned to Philadelphia and filed his report on the completed survey

1803	Published report: *The Journal of Andrew Ellicott* in Philadelphia
1803–9	Appointed secretary of the Pennsylvania Land Office
1811	After two years in private work, appointed to survey the boundary between Georgia and North Carolina
1813	Appointed as professor of mathematics at West Point Military Academy
1814	Published *The Journal* in Philadelphia
1817	Appointed to assist in the survey of the boundary between the United States and Canada
1820, August 28	Died at West Point, New York

CHAPTER I.

The author leaves Philadelphia—arrives at Pittsburgh—obtains boats, and proceeds down the Ohio River to its mouth—some account of the river, adjoining country, and inhabitants.

September 16th, 1796, I took leave of my family about ten o'clock in the morning, and proceeded to Chester and dined; then rode to Wilmington and staid all night.—Thermometer was 78° in the afternoon.

17th, Left Wilmington at half past five in the morning, breakfasted at Christiana, dined at Elkton, proceeded to the Susquehannah, crossed the ferry and lodged at Havre de Grace.—Thermometer 60° in the morning, rose to 70°, fell to 62° in the evening. Autumnal squalls and showers in the afternoon—The water in the Susquehanna was 73°.

18th, Left Havre de Grace at five in the morning, breakfasted at Hartford, dined at Baltimore, lodged at my mother's on Potapsco [Potapsco River in Maryland].—Thermometer 57° in the morning, rose to 68°.

The country from the Susquehannah to Potapsco, does not appear to be in a better state of cultivation, than it was twenty-six years ago. This disagreeable circumstance is no doubt principally owing to the system of domestic slavery, which yet continues to prevail in the southern states. Early impressions made upon the mind, and habits acquired in youth, are rarely obliterated, though condemned by daily experience.

That domestic slavery is wrong in a moral point of view is evident from the ordinary principles of justice: And that it is politically wrong may be deduced from the following facts. *First,* that a tract of country cultivated by slaves, is neither so well improved, rich or populous, as it would be if cultivated by the owners of the soil, and by freeman. *Secondly,* slaves cannot be calculated upon as adding to the strength of the community, but frequently the contrary, for reasons too obvious to detail. Notwithstanding those facts are constantly in view, they rarely produce the necessary effects upon minds early habituated to the custom of domestic slavery.[1]

19th, Remained at my mother's.—Thermometer 55° in the morning, rose to 70° in the afternoon—Water in the river 57°.

20th, About 11 o'clock in the forenoon took leave of my mother, brothers and sisters, and rode to Reister's town and got some refreshment, then proceeded about seven miles further and stayed all night.—Thermometer 51° in the morning, rose to 71°.

21st, Set out before sunrise, rode 10 miles and took breakfast, the morning was fine, and pleasant, went on to McCallester's town and dined. The town is handsome, and appears to be improving, which is not the case with Reister's town. The population of towns and villages, is generally very rapid till it becomes sufficient for the commerce of the surrounding country, and afterwards increases, or decreases with the general state of the improvement of the district, unless aided by something peculiarly favourable to its situation.

Left McCallester's town at 3 o'clock in the afternoon, and rode to Oxford, where I stayed all night.—Thermometer 53° in the morning, rose to 78°.

22nd, Left Oxford before sunrise, rode twelve miles and took breakfast; and then proceeded to Shippensburgh. On the way crossed a spur of the Blue mountain, on which peaches were uncommonly plenty, and in great perfection. Dined at Shippensburgh, where I expected to meet our commissary Mr. Anderson, but found that he had not arrived.—Thermometer 63° in the morning, fell to 53° in the evening.

23rd, Walked out about half an hour before sunrise and perceived a fine hoar frost.—The thermometer was 40° on the outside of my window; but when placed on the ground among the grass it fell to 35°; it was then placed on a fence-rail which was covered with frost and fell to 34°; but upon scraping together a small quantity of the frost, and applying it to the bulb, the mercury immediately fell to 32°, because such frosts frequently appear when

1. Ellicott's thoughts here reflect a popular view of slavery, that it: (1) is morally offensive to humanity as a whole, and (2) that it is inefficient when compared to free labor for the economic benefit of society. Congress, under the Articles of Confederation, passed the Northwest Ordinance in 1787, which included a provision that prohibited slavery north of the Ohio River. In 1790, however, Congress passed the Southwest Ordinance, which permitted slavery in the areas south of the Ohio River.

the thermometer stands 6 or 7 degrees above that point. This mistake must have arisen from supposing the degree of heat where the thermometer is suspended, and where the frost appears, to be the same; which upon experiment will be found not to be the case. After breakfast, walked out to the three large springs, being the principal sources of that fine stream of water passing through the town, and found water in each of them 51½°. In the afternoon the thermometer rose to 69°. The commissary not arriving, I set out about 3 in the afternoon, and rode to Strasburgh.—The water in the large spring in the town 51°.

24th, Left Strasburgh early in the morning. A very heavy hoar frost, vines, potato tops and corn leaves killed in the vallies between the mountains.[2] Breakfasted at Dunn's dined at Bird's and stayed all night at Wild's tavern at the foot of the Sideling Hill. Peaches in abundance along the road on the north sides of the hills. The frost the preceding night struck much more severely in the vallies, than on the mountains.—Thermometer in the morning on the outside of my window as at 35°, rose in the afternoon to 71°.

25th, Took an early breakfast, and rode to Hartley's and dined, from thence proceeded to Ward's and stayed all night.—Thermometer 35° in the morning, rose in the afternoon to 73°.

26th, Proceeded to the foot of the Alleghany mountain and breakfasted, ascended the mountain about 10 o'clock in the morning and proceeded to Stoy's town and dined, then rode to Mr. Wells' and stayed all night.—Thermometer 53° in the morning, rose to 71°, fell to 72° in the evening. The water in a good spring on the top of the mountain was 55°. The frost appeared to have fallen very partially on the mountains: in some places the vegetation was destroyed, and in others equally elevated, it was not touched.

A number of the farmers on the mountain were engaged in cutting their Buckwheat, and Oats; but Mr. Wells informed me that this was not common so late in the season. The summer on the mountain, is not sufficiently long to bring Indian corn to perfection.

27th, Left Mr. Wells' before sunrise. A very heavy hoar frost: crossed the Laurel Hill, and took breakfast at Freeman's tavern: crossed the Chestnut

2. A reminder that Ellicott was a phonetic speller throughout his writings. He also used British spellings and style conventions, and his inconsistencies in spelling (including of persons' names and place-names), grammar, capitalization, and syntax have been retained in the text. "Hoar" is an English term meaning "white" that was used at one time to denote white hair in old men, but that, of course, when used in the term "hoar frost" refers to the white icy crystals.

ridge, and dined at Baldrage's [tavern]: proceeded from thence to Greensburgh, and stayed all night.—Thermometer 35° in the morning on the outside of my window; rose to 80° after crossing the Chestnut ridge.—The water, in a good spring at the base of the Laurel hill 58°, and in a spring on the west side of the Chestnut ridge 57°. It may appear singular, that the water which falls out of the west side of those mountains, should be the warmest that was examined on the road: whether it be owing to subterranean heat, or some other cause, is yet uncertain.—The coldest water in Greensburgh was 55°.[3]

28th, Cloudy: left Greensburgh at seven o'clock in the morning, and rode to Col. John Irwin's and took breakfast, from then to McNair's and dined. Left McNair's in a heavy rain, which continued till I arrived at Pittsburgh—Thermometer 60°in the morning, rose to 68°.

The morning after my arrival at Pittsburgh I waited upon Major Craig, and found that he had two boats ready, one of them flat-bottomed, commonly called a Kentucky boat; the other a second hand keel-boat. These being insufficient he was requested to procure another, which he did in a few days, it was likewise a second hand one. After leaving Major Craig, I waited upon Col. Butler, and presented an order from the secretary of war [James McHenry] for a military escort: he gave me assurance, that the men should be ready by the time the waters were sufficiently high to descend the river. The wagons with our stores, instruments and baggage arrived on the 3rd of October.

On the 4th, I examined the state of the instruments, and found that some of them were injured by the jolting of the wagons; repaired them on the 5th. On the 6th, found the water in the Alleghany river 68°; that in the Monongahela 64°; and a few paces within the coal-pit, the temperature of the water was 51°.

On the 16th, there was a small rise of the water, and the three boats were

3. Ellicott paid careful attention to scientific observation, which he knew was of interest to his colleagues. It was his talent for preciseness and scientific accuracy that led to his appointment as the U.S. commissioner by President Washington, and even his Spanish counterparts recognized his scientific ability in these areas. Spanish governor Gayoso in Natchez later wrote a confidential letter to Prime Minister Godoy on the subject, noting: "I cannot help but exaggerate the astronomical abilities of Don Andres Ellicott, to which is joined a degree of practical experience which is unique among contemporary astronomers because of the vast number of occasions he has had to do similar operations on the boundaries of different states in the Union" (Don Manuel Gayoso de Lemos to the Principe de la Pas [Prince of Peace] June 6, 1798, trans. Jack D. L. Holmes, Archico Historico Nacional [Madrid], Seccion de Estado, legajo 3900).

sent off, but had not water to proceed more than three miles. On the 20th, Gen. [James] Wilkinson, and his family arrived, and he very politely gave his boat up to me; it was a second hand one, but the cabin was new and spacious. The 21st and 22nd, were spent in making some repairs to the boat, and on the 23rd, I went on board of it, and proceeded down the river to the others, and made such a distribution of the loading, that each vessel drew about the same water.

During my stay at Pittsburgh¸ the fogs were very heavy, every morning except two, and on those days we had rain.—Thermometer was at no time below 48° nor above 71°.

The town of Pittsburgh continues to improve, the situation is favourable, being on a point of land formed by the confluence of the Alleghany and Monongahela rivers, from which circumstance it enjoys a considerable trade.

24th, Got under way about 10:00 in the forenoon, but the water so low, that it was with difficulty we made eight miles.—Thermometer rose to 76°. The morning was very foggy.

25th, Left the shore at sunrise. The large boat was stopped for want of a sufficiency of water, three times in the course of the day; but by the exertions of about thirty men, she was brought along. Fog in the morning—Thermometer rose from 53° to 71°.

26th, Got under way early in the morning. The large boat was so much injured by dragging her over the stones, that the men had to keep ladling [baling] out the water all last night: proceeded with great difficulty down to a small town, opposite to the mouth of little beaver. The large boat did not arrive. Fog in the morning—Thermometer rose from 42° to 70°.

27th, The large boat arrived early in the morning, but so much injured that we had to unload her, stop the leaks and make some repaires; reloaded about 4 o'clock in the afternoon, and proceeded a short distance down the river, to get clear of the town, where some of our men got intoxicated, and behaved extremely ill. This will generally be found the case in all small, trifling villages, whose inhabitants are principally supported by selling liquor to the indiscreet and dissipated in the neighbourhood, and to the imprudent traveller. Fog in the morning—Thermometer rose from 41° to 69°.

28th, Left the shore early in the morning; the fog was so thick that when our boats were within twenty yards of each other, they could not be discovered by any of the persons on board. We made about sixteen miles this day. The water had but little motion.—Thermometer rose from 39° to 61°.

29th, Got under way very early in the morning: but little fog; the atmosphere had been so full of smoke ever since we left Pittsburgh that it was but seldom we could see the river distinctly, but it was carried off this morning by a smart north west wind.

The buildings on the river banks, except in the towns, are generally of the poorest kind, and the inhabitants who are commonly sellers of liquor, as dirty as their cabins, which are equally open to their children, poultry and pigs. This is generally the case in new settlements; the land being fresh, produces with little labour the immediate necessities of life, from this circumstance the habit of industry is diminished, and with it the habit of cleanliness.

Encamped in the evening opposite to the Mingo bottom which is rendered memorable for the inhuman murder of the Indians of that name, who resided on it, either by, or at the instigation of Capt. Cresup, Harmon Greathouse, and a few others. This outrage was followed by a war of retaliation, which continued for many years with a cruelty scarcely to be equaled in the annals of history. [Present location is Mingo Junction, Ohio, a historic site from which American militia in 1782 began their extermination of the Mingo Indians, and others presumed to be hostile.]

The evening became calm, and the atmosphere again loaded with smoke, occasioned by the dead leaves and grass, over a vast extent of the country being on fire, which during the night, illuminated the clouds of smoke and produced a variegated appearance beautiful beyond description. Our smokey weather in spring and autumn, is probably the effect of fires extending over the vast forests of our country.

Our people were much fatigued by dragging our boats over the shoals.—Thermometer rose from 39° to 57°.

30th, Detained till one o'clock in the afternoon by the commissary who was endeavouring to procure some meat; but being disappointed we proceeded down the river to Buffalo, where we were again disappointed. Buffalo is a decent village, and is situated on the east side of the river, just above the mouth of a rivulet of the same name. Left Buffalo in the evening, and proceeded about three miles and encamped. The morning was very smokey.—The thermometer rose from 30° to 51°.

31st, Our commissary went on a few miles before us, and purchased three beeves [beef], we followed between seven and eight in the morning; it was then so smokey, that we proceeded with difficulty. Encamped about four miles above Wheeling. Several of our men were indisposed with sore throats, owing probably to colds contracted from their frequent wettings.—Thermometer rose from 36° to 47°.

November 1st, About one o'clock in the morning we had a furious gale of wind; it appeared, as if nature was making an exertion to free the atmosphere from the astonishing quantity of smoke, with which it had been filled for many days. Stopped at Wheeling and took the latitude, and then proceeded to the mouth of Grave Creek and encamped. Went to view the amazing monuments of earth, thrown up many ages ago by the aborigines of the country, for some purpose unknown to us. One of those monuments [Grave Creek Mound] is more than 70 feet high: it has a cavity or depression on the top, in which a large oak tree was growing.[4] The atmosphere again became smoky in the evening.—Thermometer rose from 30° to 52°.

2nd, Found that one of our soldiers had deserted after being detected in stealing liquor: made search for him but to no purpose. Got under way at ten o'clock, but our progress was much impeded by dragging our boats over the shoals. Cloudy with an appearance of rain most of the day, but cleared off in the evening. Thermometer rose from 29° to 50°.

3rd, Got under way about 7 o'clock in the morning, and continued down the river till sun down. The large boat was impeded by a strong head wind, and did not overtake us till eight o'clock in the evening. Cloudy in the morning but no fog, clear night.—Thermometer rose from 36° to 55°.

4th, Set out early in the morning, but our progress was impeded by head winds. Encamped at sunset. The large boat did not overtake us. Cloudy with thick smoke all day.—Thermometer rose from 35° to 56°.

5th, Left the shore before sunrise, and proceeded down the long reach: at the lower end of it the water was so shallow, that we were two hours employed in dragging our boats over the gravel, and then encamped. The large boat still behind. The fog was so thick in the morning, that for four hours when in the middle of the river, we could see neither shore; some appearance of rain in the evening;—Thermometer rose from 33° to 49°.

6th, Left the shore at seven o'clock in the morning.—Thermometer rose from 25° to 56°.

7th, Set off at sunrise, and arrived at Marietta about eleven o'clock in the forenoon. Unloaded the boats to stop the leaks, and make some repairs. Smoke as in the morning;—Thermometer rose from 25° to 56°.

4. Grave Creek Mound is located in present-day Moundsville, West Virginia, and was built during the Adena culture of the Woodland Period, ca. 1000 BCE.

8th, The men were employed in repairing the boats. Viewed the amazing works thrown up many ages ago by the Indians. They are the most regular of any I have seen. Some smoke and fog in the morning—Thermometer rose from 31° to 52°.

9th, Our men still employed in repairing the boats. Our large boat arrived in the evening. The smoke was so thick all day that could not see over the river. Thermometer rose from 34° to 53°.

10th, The boats were repaired, and loaded by one o'clock in the afternoon, we then proceeded down the river.

Marietta is a handsome town, standing on a high bank, on the west side of the Ohio river, just above the mouth of the Muskingum. The annual rise of the water has sometimes inundated the lower part of the town. The latitude by a mean of four good observations appeared to be 39°24′21″.

During our stay, we were treated with great politeness by Col. Sproat, and a sensible young gentleman, by the name of Tupper, a son to the General of that name. I paid a visit to Gen. Putnam [Rufus Putnam, first surveyor general of the United States, 1796–1803], who had lately been appointed Surveyor General of the United States, and presented him with one of my pamphlets upon the variation of the magnetic needle. He did not return the visit, neither did I hear from him afterwards. The fog and smoke was so thick till ten o'clock in the forenoon, that we could not distinguish one person from another the length of one of our boats. Encamped at sun down;—Thermometer rose from 34° to 56°.

11th, Left the shore at seven o'clock in the morning. Passed the little Kanhawa, and afterwards a miserable village by the name of Belle Prae, next a floating mill, and lastly, the mouth at Hockhocking.

The ordinary streams of water in that part of the western country, so universally fail in the summer, and beginning of autumn, that the inhabitants are under the necessity of having recourse to floating mills, or to others driven by the wind, or worked by horses to grind their corn. Those floating mills are erected upon two, or more, large canoes or boats, and anchored out in a strong current. The float-boards of the water wheels, dip their whole breadth into the stream; by which they are propelled forward and give motion to the whole machinery. When the waters rise, and set the other mills to work, the floating ones are towed into a safe harbor, where they remain till the next season. Although floating mills are far inferior to permanent ones driven by water, they are nevertheless more to be depended upon than wind mills, and may be considered as preferable to those worked by horses. The lessening of manual labour and that of domestic animals, is a subject at all times, and in all countries, which merits the attention of the moralist, the philosopher and legislator. The effect produced by either the wind, or water, is not attended with any expense, and while those elements are directed to the execution of some valuable purpose, manual labour, and that of domestic animals, may be employed in a manner equally beneficial to the community, and more to their ease, safety, and convenience.

Encamped opposite to a miserable village called Belle Ville: made 24 miles this day. Fog in the morning, and smoky all day. Thermometer rose from 37° to 66°—Water in the river 45°.

12th, Left the shore at daylight. Dragged our boats over several shoals, and encamped at sundown. Very smoky all day.—Thermometer rose from 37° to 55°.

13th, Got under way very early; had to drag our boats a considerable distance over the shoals, with which one of them was much injured. Encamped about sun down and caught a number of fine cat fish, one of them weighed more than 48 pounds. Very smoky all day.—Thermometer rose from 27° to 52°.

14th, Were under way at daylight, and arrived in the evening at Point Pleasant, a small and indifferent village on the east side of the river, just above the mouth of the great Kanhawa [Kanawha River]. We were politely treated by Mr. Allen Proyer of that place. Near to where the village now stands, was fought the memorable battle between a detachment of Virginia militia, (commanded by Col. Lewis,) and the Shawnee and Delaware Indians. The engagement continued several hours, and the victory was a long time doubtful, and alternately appeared to favour each party; courage, address and dexterity equally characterized both; but the Virginians remained masters of the field.

We had a light shower of rain in the morning, and two in the evening. The atmosphere yet filled with smoke.—Thermometer rose from 41° to 62°.

15th, Arrived at Gallipolis about eleven o'clock in the forenoon. This village is situated on a fine high bank, on the west side of the river, and inhabited by a number of miserable French families. Many of the inhabitants that season fell victims to the yellow fever, which certainly originated in that place, and was produced by the filthiness of the inhabitants, and an unusual quantity of animal, and vegetable putrefaction in a number of small ponds, and marshes, within the village. Of all the places I have yet beheld, this was the most miserable.

There are several Indian mounds of earth, or barrows, within the vicinity of the village. Detained the remainder of the day in procuring meat. The smoke yet continues.—Thermometer rose from 40° to 62°.

16th, Left Gallipolis at eight o'clock in the morning. About two in the afternoon, one of our boats went to pieces in a body of strong rough water, and it was with difficulty conveyed to the shore without sinking. Encamped at sunset. A strong north wind all day, which carried off the smoke.—Thermometer rose from 37° to 52°.

17th, Left the shore at daylight: got fast on a shoal, which extends across the river when it is low, by which we were detained two hours in drawing our boats over it: detained again from one till two o'clock in the afternoon in passing another shoal. Shortly afterwards, two black men belonging to a boat we fell in with in the morning, were drowned in attempting to go to the shore in a canoe. We sent some of our people to aid in finding them, one was taken up but could not be revived. Light showers of rain in the evening.—Thermometer rose from 27° to 63°.

18th, Left the shore before sunrise, got fast, and were detained for more than one hour. Passed the mouth of Sandy Creek, which is one of the boundaries between the states of Virginia and Kentucky. Encamped just before dark. A heavy fog in the morning, the afternoon remarkably clear.—Thermometer rose from 43° to 64°. Water in the river 46°.

19th, Set off before day, got fast on a log where we remained till near sunrise. The fog was so thick, that we could neither discover sand-bars nor logs, till it was too late to avoid them. The fog disappeared about ten o'clock in the forenoon: reached the mouth of the Big Scioto a short time before noon, and took the sun's meridional altitude, by which the latitude speared to be at 38°43′28″ N. [The meridian is the north-south line of the observer in the horizon coordinate system by which to determine latitude.] The waters of the Scioto have a strong petrifying quality. We collected several fine specimens. Proceeded down the river, and encamped after sunset.

Cloudy, with thunder and lightning in the evening, accompanied by a shower of rain. The atmosphere again filled with smoke.—Thermometer rose from 43° to 65°.

20th, Left the shore at daylight, had a sharp thunder gust between six and seven o'clock, in the morning. At ten o'clock, I left the boats, and went on shore with my skiff to view the salt works, which are about one mile from the river, in the state of Kentucky, and collected the following information, viz. that

300 gallons of water, produce one bushel of salt: they had 170 iron kettles, and made about 50 bushels of salt per day, which sold for 2 dollars cash per bushel, or 3 dollars in trade, as they term it. The salt lick, or spring, is situated in the bed of a small creek, which when high overflows it. The backwater from the river also inundates the lick, or spring, when high, together with all the works. From these causes, they were not able to carry on the business more than eight months in the year. But the greatest difficulty they found, was with the bitter water, which I supposed was what the manufacturers of salt in England, call bittern, and drains from the salt after it is granulated, and stowed away in the drales [vats] to dry. However I asked the manager for an explanation who replied, 'Bitter water is mixed with the salt water, and separated from it in this manner. After boiling the water, till the salt will just begin to crystallize, it is ladled out into troughs, or vats, and let stand for some time, and the bitter water which assumes a dark brown colour, floats on the top of the salt water, and is skimmed off, and thrown away, and if this separation was not made, the salt would not properly granulate, but become tough, and form a hard mass of a bitter taste, and quite useless: and if the bitter water was evaporated, it would leave a hard mass of matter, of a disagreeable, and nauseous taste, but as it was useless, they had none of by them.' He likewise informed me, that the bitter water was very injurious to cattle, by inflaming the skins of such as frequently drank it, that the same effect was often experience by their workmen, and that in a few hours it would destroy the quality of leather. The lick, or spring, does not appear to have been much frequented either by buffaloes, or deer, and the cattle in the neighbourhood are not remarkably fond of it.—Temperature of the water in the spring was 60°.

Returned to the river, and followed the boats, passed two villages, one in Kentucky by the name of Preston, the name of the other I did not learn. Overtook my company about sunset, and encamped. Cloudy in the evening.—Thermometer from 45° to 70°. Water in the river 50°.

21st, Set off before day, and arrived at Limestone about ten o'clock in the forenoon. It is a miserable village; left it in about an hour. Encamped at dark. Cloudy, with mist, and light showers all day.—Thermometer rose from 39° to 42°.

22nd, Rain, and hail from four till ten o'clock in the forenoon, when a heavy fall of snow began, and continued till night. The weather was so extremely bad, that we could not proceed.—Thermometer was 42° in the morning, 33° at noon, and 25° at eight o'clock in the evening.

23rd, Clear morning, and hard frost, cleared our boats of ice and snow by ten o'clock in the forenoon, and proceeded down the river. Encamped at sunset.—Thermometer rose from 19° to 27°.

24th, Got under way at eight o'clock in the morning. About eleven o'clock in the forenoon my boat struck the root of a lodged tree in the river, and was so much injured by it, that we had to put to shore and stop the leak, which detained us till noon. The river was much lower than it had ever been known, since the first settlements commenced in that country, and it was with the greatest difficulty we made any progress on that day, being obliged to drag our boats over several shoals of considerable extent, the weather at the time being so cold, that the men's clothes froze stiff almost as soon as they came out of the water. Encamped after sun set on the east side of the river, opposite to a village called Columbia.—Thermometer rose from 14° to 25° Water in the river 40°.

25th, Proceeded to Cincinnati, where we arrived about ten o'clock in the forenoon, and found ourselves under the necessity of procuring another boat, in place of one which was rendered useless by dragging it over rocks, stones, and shoals and repairing the one I had from General Wilkinson. The waters were so low that no boats but ours had reached that place from Pittsburgh since the preceding August, and the season was then so far advanced that no others could be reasonably expected. Our success was owing to the number of people we had with us, and whose quiet submission to unusual hardships does them great credit. A clear day.—Thermometer rose from 20° to 25° and fell to 14° at nine o'clock in the evening.

26th, A clear day.—Thermometer 3° below 0 at sunrise, rose to 21° but fell to 9° at ten o'clock in the evening. Water in the river 33½°.

27th, A clear day.—Thermometer 7° at sunrise, rose to 25°. The water rose this day about 2 inches.

28, A little fine snow last night.—Thermometer rose from 15° to 37°. In the evening the water appeared to have risen about one foot.

29, A fine pleasant day, and the snow began to melt. The boat being repaired, and another procured, we left Cincinnati in the evening. The water had risen about three feet.—Thermometer rose from 35° to 46°.

Cincinnati was at that time, the capital of the North Western Territory: it is situated on a fine high bank, and for the time it has been building, it is a very respectable place. The latitude by a mean of three good observations is 39°5′54″ N. During our stay, we were politely treated by Mr. Winthrop

Sargent, secretary of the government, and Captain Harrison who commanded at Fort Washington.[5]

30th, Floated all night, and passed the mouth of the Great Miami early in the morning. The river in much better order for boating. Rain and snow all day. Encamped in the evening.—thermometer rose from 26° to 37°. Water in the river 35°.

December 1st, Left the shore at daylight. Snow and rain the whole day. Encamped about dark.—Thermometer 37° all day.

2nd, Set off at daylight. Stopped and took breakfast at a small village just above the mouth of Kentucky river. Made but little way on account of a strong head wind. Encamped just before dark. Cloudy all day.—Thermometer continued at 29°.

3rd Got under way at daylight; but made little way on account of head-winds. Encamped at dark.—Thermometer rose from 23° to 30°. Water in the river 33°.

4th, Set off at daylight, and proceeded down to the rapids, and encamped. The water was so low that the pilots would not be answerable for the safety of the boats in passing the falls.—Thermometer rose from 28° to 35°.

5th, Concluded to risk the boats, rather than be detained till the rise of the waters, and sent all the people, who were either afraid of the consequences, or could not swim, round by land, and a little after noon, all the boats were over, but not without being considerably damaged: the one that I was in had nine of her timbers broken. Squalls of rain and snow, all the afternoon. Thermometer rose from 24° to 29°.

6th, Spent at work upon our boats. Squalls of snow all day.—Thermometer rose from 21°to 28°.

7th, Finished repairing our boats. Cloudy great part of the day.—Thermometer rose from 18° to 26°.

8th, Detained till evening by our commissary, who was employed in procuring provisions. Set off about sun down.

The town of Louis Ville stands a short distance above the rapids on the

5. Winthrop Sargent (1753–1820) was a Federalist who was appointed as the first American governor of the new Mississippi Territory in 1798. For his later problems in Natchez, see Holmes, *Gayoso*, 84, 156, 256–59. See also the comprehensive history explored by Robert V. Haynes, *The Mississippi Territory and the Southwest Frontier, 1795-1817* (Lexington: University Press of Kentucky, 2010).

east side of the river. The situation is handsome, but said to be unhealthy. The town has improved but little for some years past. The rapids are occasioned by the water falling from one horizontal stratum of lime-stone, to another; in some places the fall is perpendicular, but the main body of the water when the river is low, runs along a channel of a tolerably regular slope, which has been through length of time worn in the rock. In the spring when the river is full, the rapids are scarcely perceptible, and boats descend without difficulty or danger.—Thermometer rose from 22° to 29°.

9th, Floated all night. Stopped in the morning to cook some victuals, and then proceeded on till sunset and encamped.—thermometer rose from 27° to 35°. Water in the river 33°.

10th, Left the shore at sunrise. About nine o'clock in the morning discovered a Kentucky boat fast upon a log, and upon examination found that it was deserted, and suspected that the crew were on shore in distress, which we soon found to be the case. The crew consisted of several men, women, and children, who left the boat two days before in a small canoe when they found their strength insufficient to get her off. They were without any shelter, to defend them from the inclemency of the weather, and it was then snowing very fast. We spent two hours in getting the boat off, and taking it to the shore, where we received the thanks of the unfortunate crew, and left them to pursue their journey.

Having a desire to determine the geographical position of the confluence of the Ohio and Mississippi rivers, and the large store boat not being calculated, I left her with directions to follow with all possible dispatch, and pushed on myself for the mouth of the river. Stopped at sun down, to give our men time to cook some victuals: set off at eight o'clock in the evening, and proceeded down the river against a strong head wind till almost midnight, when it became so violent that we had to put to shore. Snow great part of the day.—Thermometer rose from 21° to 28°. Water in the river 33°.

11th, Left the shore at daylight, and worked against a strong head wind till sunset, then went on shore to dress some victuals. Cloudy great part of the day.—Thermometer rose from 25° to 29°. Left the shore at eight o'clock in the evening, and worked all night against a strong head wind.

12th, This morning the country appeared much lower, the hills or mountains smaller, and more pyramidical. No snow to be seen. The day fine and clear.—Thermometer rose from 32° to 45°. Put to shore at sunset to cook. Got underway before seven o'clock and proceeded all night.

13th, Went to the shore at eight o'clock in the morning to cook. Set off in two hours. Snow and rain till evening. Stopped at sun down to cook. Set off at eight o'clock, and floated all night. The country flat, and a large portion of the river banks subject to inundation. Passed Red Bank about nine o'clock in the evening.—Thermometer rose from 33° to 37°.

14th, The country yet continues flat. Put to the shore about eight o'clock in the morning to dress some victuals. Set off in two hours. Passed the Wabash about nine o'clock in the evening.—Thermometer rose from 28° to 37°. Water in the river just above the freezing point.

13th, Much ice in the river. Stopped at an Indian camp, and procured some meat. Dined at the great cave. This cave may be considered as one of the greatest natural curiosities on the river, and I have constantly lamented that I could not spare time to make a drawing of it, and take its dimensions. It is situated on the west side of the river. The entrance is large and spacious, and remarkably uniform, the dome is elliptical, and the uniformity continues to its termination in the hill.

Stopped about sunset to take in some wood. Set off in half an hour and floated all night. Cloudy part of the day.—Thermometer rose from 21° to 41°.

16th, At eight o'clock in the morning, one of our boats unfortunately ran on the roots of a tree, which were under water, and bilged. We spent till near one o'clock in the afternoon in repairing her, and then proceeded down the river till about sunset and encamped. The weather that day was very pleasant.—Thermometer rose from 35° to 51°. Passed Cumberland river at 3 o'clock in the afternoon.

17th, Set off at day light, and about nine o'clock in the morning passed the mouth of the Tennessee, and in two hours afterwards arrived at Fort Massac, and was politely received by the commandant Captain [Zebulon] Pike, who with the surgeon doctor Hammel dined with me. The fort stands on a high bank, on the west side of the river. The stones appear to be composed of ferruginous matter and gravel. A considerable quantity of land both above, and below the fort, is annually inundated.

Left the fort about two o'clock in the afternoon, and proceeded down the river till dark, and then came to. Thunder and lightning in the evening.—Thermometer rose from 48° to 64°. Water in the river 36°.

18th, Left the shore at daylight, and arrived at the mouth of the Ohio about two o'clock in the afternoon.—Thermometer 46° at sunrise; but fell to 24° in the evening. Water in the river 34°.

19th, Set up a clock, and prepared to make some astronomical observations for the purpose of determining the latitude and longitude of the confluence of those great, and important rivers: for those, and the thermometrical observations made at this place, see the Appendix.

The map of the Ohio river which accompanies this work, is laid down from the best materials I could procure, a number of the latitudes between Pittsburgh and rapids, were taken by myself: from thence down to the Mississippi, the latest charts have been used, except in a few places which have been corrected by my friend Don Jon Joaquin de Ferrer, an ingenious Spanish astronomer. The map is divided into two parts that it may not be too large to fold in a quarto volume, and at the same time of such a size, as to shew distinctly the errors that may hereafter be discovered and serve as a basis for future corrections.

The Ohio river, is formed by the junction of the Allegany and Monongahela rivers, at Pittsburgh, which name it retains till it falls into the Mississippi. It may not be improper here to observe, that all the Indians residing on the Allegany, ever since my acquaintance with the western country have called that branch, as well as the main river, the Ohio, and appeared to know it by no other name.

The Ohio is certainly one of the finest rivers within the United States, whether considered as to magnitude, the great extent of its course, or the outlet it affords to an immense and fertile country rapidly filling with inhabitants.

The bottom and sides of the river are stony, from Pittsburgh down to the low country, which is generally supposed to be about eight hundred miles. The strata of stone are horizontally disposed, and principally consist of either freestone, or limestone. This horizontal disposition of the strata of stone is observable through a very large extent of the United States. I have traced it from Oswego, up Lakes Ontario and Erie, with all the waters falling into them, and through all the western parts of Pennsylvania, and down the Ohio, wherever hills or mountains are to be seen.

The flat, or bottom lands on the Ohio, are not surpassed by any in the United States for fertility; but in many places they are small, and inconsiderable; being limited by hills or mountains, on one side, and the river on the other. A large proportion of the hills and mountains, are unfit for agricultural purposes, being either too steep, or faced with rocks. The hills and mountains on the east side of the river, generally increase in magnitude, till they unite with the great ridge, commonly called the Allegany: but on the west side they decrease, till the country becomes almost a dead level.

The country produces all the immediate necessaries of life in abundance, and far beyond the present consumption of the inhabitants; the residue, with many other articles, such as hemp, cordage, hard-ware, some glass, whisky, apples, cider, and salted provisions, are annually carried down the river to New Orleans, where they find a ready market. Mines of pit coal (lithanthrax) are not only abundant, but inexhaustible from Pittsburgh many miles down the river.

The inhabitants of no part of the United States are so much interested in establishing manufactories, as is this [area]. They possess the raw materials, and export their produce with ease, but their imports are attended with difficulty, great risk, and expense. And so long as they receive neither bounties, nor uncommon prices for their articles of exportation, and depend upon the Atlantic states for their supplies of European manufactures, the balance of trade will be constantly against them, and draw off that money, which should be applied to the improvement of the country, and payment of their taxes. To this source, may some degree be traced, the character the inhabitants have too generally had bestowed upon them of insurgents, and disorganizers; to a few individuals these epithets may be applied, but not to the body of the people. In order to judge fairly on this question, it will be necessary to take into view the local situation of the inhabitants. In the Atlantic states every article however minute, if a necessary of life, will not only find a ready market, but command cash. On the Ohio, and its waters, almost the only article, which heretofore found a ready market at home, and would command cash, was their own distilled spirits. The taxing of this article would therefore be but little different from taxing every article in the Atlantic states, which commanded cash. Such a tax as the latter, I am inclined to believe, would be collected with difficulty, and probably with the same propriety, given the same turbulent character to a great majority of the nation.[6]

6. Ellicott's reference here is to the famous "Whiskey Rebellion" (1794) among inhabitants of western Pennsylvania, on the frontier, with which he was familiar. Governor-General Baron de Carondelet and Governor Gayoso interpreted the 1794 rebellion as an example of the opposition in the West and South to the Federalist regime in Philadelphia, and along with the intrigues from General Wilkinson and Senator Blount as well as others, led them to believe it might be possible to detach the western settlements from the United States and thereby to

I am far from justifying any opposition by force, to the execution of laws constitutionally enacted, they must either expire, or be constitutionally repealed; a contrary proceeding must terminate in the destruction of all order, and regular government, and leave the nation in a state of nature: but at the same time, it is a duty incumbent on the legislature, to attend to the local situations of the several constituent, or component parts of the union, and not pass laws, which feebly felt in one part, and be oppressive in another. That some turbulent persons are to be met with on our frontiers, every person possessed of understanding and reflection, must be sensible, will be the case so long as we have a frontier, and men are able to fly from justice, or their creditors; but there are few settlements so unfortunate as to merit a general bad character from this class of inhabitants.

The people who reside on the Ohio and its waters, are brave, enterprising, and warlike, which will generally be found the strongest characteristic marks of the inhabitants of all our new settlements. It arises from their situation; being constantly in danger from the Indians, they are habituated to alarms, and acts of bravery become a duty they owe to themselves, and to their friends. But this bravery, too frequently when not checked by education, and a correct mode of thinking, degenerates into ferocity.

Vessels proper for the West India trade, may be advantageously built on the Ohio, and taken with a cargo every annual rise of the waters down to New Orleans, or out to the islands. The experiment has already been made, and attended with success.

That the climate on the Ohio, does not appear to be inferior to that of any part of the Union. The inhabitants enjoy as much health, as they do on any of the large rivers in the Atlantic states. At Pittsburgh, and for a considerable distance down the river, bilious complaints are scarcely known; but they are frequent at Cincinnati, and still more so at Louisville near the rapids.

The timber growing on the river bottoms, is with a few exceptions, such as grows on the east side of the mountains in the middle states, *viz.* Black walnut (juglans nigra,) butter nut, (juglans cinerea of wengemheim,) juglans hickory of several species. Sugar maple, (acer saccharinum,) button wood, or sycamore, (plantanus occidentalis,) tulip tree, (liriodendron tulipifera,)

---

the interests of Spain. The passage of the Hamiltonian excise tax in 1791 on distilled spirits affected almost exclusively the frontiersmen, and was one more reason, along with foreclosures, banking, monetary issues, and trade policies, that they felt discriminated against by the northeastern merchants and their political (Federalist) allies.

will berry, (prunus Virginiana,) papaw, (anona triloba,) spice wood, (laurus benzoin,) black willow, (salix nigra,) ash, (fraxinus Americana,) elm, (ulmus Americana,) mulberry (morus rubra,) locust, (robinia pseud-acacia,) honey locust, (gleditsia triacanthus,) buck eye, (aesculus flava,) beech, (fagus ferruginea). To which might be added many more of less note and value. The uplands likewise produce many of the foregoing, with a variety of others, such as black oak, (quercus nigra,) white oak, (quercus alba,) red oak, (quercus rubra,) chestnut, (fagus American,) dog wood (cornus Florida,) cucumber tree, (magnolia acuminate,) and sassafras, (laurus sassafras.) [Ellicott was a scientist, and his descriptions of plant life in the region were seen as part of his scientific obligation and contribution to knowledge of the region. He, like the other members of the American Philosophical Society, was familiar with the scientific writings of their European colleagues, such as the Swedish botanist Carl Linnaeus (1707–1778), whose published works from 1735 to 1753 established his international reputation for the development of his taxonomy system and of plant classification. His two-volume study *Species plantarum* (1753) was widely regarded as definitive for plant classification.]

CHAPTER II.

Occurrences at the mouth of the Ohio, and down the Mississippi to the town of Natchez.

The day after we arrived at the confluence of the Ohio and Mississippi rivers, they were both so full of ice that it would have been impossible to navigate either of them with any degree of safety; and the day following, each appeared like a vast mass of ice and snow in motion.

On the 22nd, both rivers closed, and made a romantic appearance, from the piles of ice which were thrown up in a variety of positions.

We now became alarmed for the fate of our store boat, which we left behind on the 11th, and otherwise found our situation very disagreeable; not expecting to be overtaken by such extreme cold weather, we were not prepared to meet it. Great part of our blankets and stores, flour excepted, were behind. Our instruments, baggage, and other articles, were all taken to the top of the bank, on the east side of the river, where we encamped: and for a number of days the cold was so intense, that we had to keep up large fires both day and night, to prevent our being frozen.

In consequence of the severity of the frost, the water in the river had fallen during the first four days after our arrival about four feet, which left our boats frozen so fast in the mud, and loaded with such a quantity of ice, that it would have been impossible to save them, if a sudden rise of the water had taken place; and the loss of our boats, we were sensible, would involve us in very serious difficulties, as a considerable tract of country where we were encamped, was inundated every annual rise of the waters, and this rise of the waters might be expected with the breaking up of the ice.

On the 23rd, we sent three men up the river, to obtain if possible, some intelligence respecting our store boat.

On the 25th, about eight o'clock in the evening, the ice gave way, and continued to move about fifteen minutes. We had labored hard on that and the preceding days to free the boats from ice, and loosen their bottoms from the frozen mud and sand; and every person was now engaged to save them. Just as the ice gave way the water rose about four feet, and one of the boats which had not sunk very deep into the mud, rose with it, and was immediately hauled out on the beach; but the other continued fast until loosened by the ice, and was carried away about 80 yards; but the ice stopped so quick, that she received no material injury. A passage was immediately opened through the ice to the boat, which was then drawn out of the water, and safely landed with the other.

The next morning about sunrise the ice gave way again, and notwithstanding the extreme degree of cold, it continued to move the whole day in so great a mass, that the water was not to be seen. Both rivers made the same appearance, and as our boats were now safe, we were enabled to contemplate the prospect which was grand and awful, with some degree of pleasure and composure. The concussion of the ice at the junction of the two rivers produced a constant, rumbling noise, for many hours, similar to that of an earthquake.

After this time, neither of the rivers was completely closed at their confluence: but a few miles above our camp, the Ohio continued shut till about the 20th of January following.

On the 2nd of January 1797, the party which had been sent up the river returned, after going as high as fort Massac [Illinois]; but could obtain no intelligence respecting our store boat. But on the 6th our anxiety was in some degree removed by the arrival of Messrs. Ellicott, junr., Rankin, and one of the labourers, who had left the boat, notwithstanding the uncommon severity of the weather, and proceeded down the river, sometimes on the ice, and sometimes on the land, till they fell in with us. By then we were informed, that they continued floating in the boat among the ice so long as it was practicable, and after running several risks, they made the land near the mouth of the Wabash, where they landed the stores, and the whole party (being about twenty in number) encamped. They likewise assured us, that every precaution they could devise, had been taken to prevent the loss of the boat, when the ice should give way.

This journey, undertaken and performed by such young men, at such an inclement season, does them great credit, and shews a spirit of enterprise, and a voluntary surrender of every comfort seldom to be met with, in committing themselves, (with what they could carry on their backs,) to the wilderness, to perform what they supposed to be a duty.

During the winter season, until the inundation commences, a number of Indians hunt and reside in the swamp; and it happened fortunately for us, that several companies of them were encamped in our neighbourhood, and by whom we were supplied with meat in exchange for flour: but exclusive of what we obtained from them, our men were not unsuccessful, and besides other game, took a great number of raccoons, and opossums, which after being skinned, and hung out in the frost a few days, afforded a pleasant kind of aliment. They appeared to abound particularly in the swamp.

A few days after we had encamped at the confluence of the rivers, Mr. Philip Nolan, so well known for his athletic exertions, and dexterity in taking wild horses stopped at our camp on his way from New Madrid [Missouri] to fort Massac [American fort on the Illinois side of the Ohio River], having two boats at the latter place shut up by the ice.[7] From him, I obtained much useful information relative to the situations, and characters, of the principal inhabitants of Natchez; which at that time was a matter of mere curiosity, but which eventually I found extremely useful.

Being pleased with his conversation, and finding that he had very extensive knowledge of that country, particularly Louisiana, I requested the pleasure of his company down the river, as we were unacquainted with the navigation of it, to which he agreed. After staying with us one night he proceeded up to [Fort] Massac, and remained there till our store boat reached that place, and accompanied her, with his two boats down to us. While in our camp he observed a number of Indians, who were from the west side of the Mississippi, and spoke to them in the several languages with which he was acquainted, but they could not understand him; he then addressed them by signs, to

7. Philip Nolan, a friend of James Wilkinson, was known as an excellent horseman. Although he dealt regularly with the Spaniards, he spoke ill of them to the Americans, including Ellicott. He was later killed in a gun battle in 1801 with Spanish troops in Texas. See Maurine T. Wilson and Jack Jackson, *Philip Nolan and Texas Expeditions to the Unknown Land, 1791–1801* (Waco, TX: Texan Press, 1988). New Madrid, Missouri, was visited also by the English merchant Francis Baily and described in his *Journal* (135–39). The original site was later destroyed by an earthquake and the flooding of the Mississippi River in 1811.

which they immediately replied, and conversed for some time with apparent ease, and satisfaction. This was the first time I had either seen, or heard of this curious language, and being led by curiosity to speak to Mr. Nolan upon the subject, he informed me that it was used by many nations on the west side of the Mississippi, who could only be understood by each other in that way, and that it was commonly made use of in transacting their national concerns. A vocabulary of part of this curious language, has been sent on to the American Philosophical Society by William Dunbar, Esq. [1750–1810] of the Mississippi Territory, and contains a much more particular account of it than I could give.[8]

About the 16th of January the weather began to moderate, and both rivers rose gradually, and carried off the ice almost imperceptibly.

On the 29th, in the evening the store boat, with Mr. Nolan's two boats arrived.

The 30th was spent in loading our boats, and preparing to leave our encampment.

On the top of the stump of a large tree, to which the zenith sector was fixed, a plate of lead was laid, containing the latitude, and longitude of that place. The stump was then covered by a mound of earth of considerable magnitude; but which will probably be demolished in a few years by the annual inundations.

31st, About 11 o'clock in the morning got under way, but the weather being bad, attended with a remarkably heavy rain, we had to encamp before sunset.

February 1st, Left the shore at daylight, and proceeded down the river to the station of one of the Spanish gallies [armed flat-bottom boats]; the master

8. William Dunbar (1750–1810) was a scientist of the first order, and later he was admitted to membership in the American Philosophical Society. He was also a local planter, an explorer, and a surveyor for the Spanish. He served on the Spanish Boundary Survey for a short period of time but resigned because of ill health and what he regarded as a conflict of interest by serving both Spain and the United States. Ellicott spoke highly of Dunbar in his writings. See the excellent analysis of him by Arthur H. DeRosier Jr., *William Dunbar: Scientific Pioneer of the Old Southwest* (Lexington: University of Kentucky Press, 2007). Dunbar was not only an excellent scientist but also an explorer. After the Louisiana Purchase in 1803, President Jefferson commissioned William Dunbar and George Hunter to explore the southern portion of Louisiana at the same time that Meriwether Lewis and William Clark were exploring the northern portion in 1804. See Trey Berry, Pam Beasley, and Jeanne Clements, eds., *The Forgotten Expedition, 1804–1805: The Louisiana Purchase Journals of Dunbar and Hunter* (Baton Rouge: Louisiana State University Press, 2014).

behaved very politely, but informed us that it would be proper to remain at his station till the next morning.

2nd, After taking some coffee with the commandant of the galley, we proceeded down to New Madrid, where we arrived about eleven o'clock in the forenoon, and were saluted on our landing by a discharge of the artillery from the fort, and otherwise treated with the greatest respect and attention by the commandant [Colonel Carlos de Hault de Lassus] and officers of the garrison.

In the evening, the commandant told me, that he had a communication to make, and for some reasons, which he did not detail, requested me to continue there two or three days. He was told, that we had been so long detained, for the want of sufficiency of water in the Ohio river to enable us to descend it with expedition, and by being so long shut up by the ice, that I could not permit myself to think of it. He then desired me to breakfast and dine with him the next day, to which I consented.

3d, After taking breakfast with the commandant, I walked about two miles out to the plains, and high lands: the soil did not appear to be of the first quality. Between the town, and plains is a swamp through which a body of water passes when the river is full. Returned about half an hour before noon, and took the sun's meridional altitude.

Immediately after dinner, the Commandant desired me to walk into a private room with him: after being seated a few minutes, he called in a clergyman of the Church of Rome, a native of Ireland, of the name of Maxwell, a well informed liberal gentleman, who acted as interpreter on that occasion. The Commandant addressed himself to me nearly as follows. 'Sir, I find myself somewhat delicately situated in consequence of your arrival, which you will readily perceive by this letter, which I received last November, from the Governor General the Baron de Corondolet.'[9] He then handed the letter to Mr. Maxwell, who read and interpreted it. It contained an order to the

9. Francisco Luis Hector, Baron de Carondelet (1748–1807), served as governor-general of Louisiana from 1791 to 1797 and was generally regarded by non-Spanish observers as one of the better Spanish officials. After his tour of duty in New Orleans, he was promoted to president of the Real Audiencia de Quito, present-day Ecuador, where he served from 1797 to 1807. For commentary on Carondelet, see Holmes, *Gayoso,* 67, 87–88, 162, 165–66, 175–76, 178–80, 197; Baily, *Journal,* 154, 157, 163; and James Pitot, *Observations on the Colony of Louisiana from 1796 to 1802,* ed. and with foreword by Robert D. Bush (Baton Rouge: Louisiana State University Press, 1979), 10–14, 40–41, 45, 48–49.

commandant, not to permit us to descend the river till the posts were evacuated, which could not [be] effected until the waters should rise. The order appeared extraordinary to me, and I believe not less so to the Commandant, as the state of the water in the Mississippi below the mouth of the Ohio, is at all times such, that no difficulty could arise in the evacuation of any of the posts claimed by the United States. In our conversation upon the subject I endeavoured to convince the Commandant, that our descending the river could not be construed into a violation of his orders. *First,* because if the want of water was an objection, when the Governor General issued the order, it was now done away by the commencement of the inundation. *Secondly,* that the Governor's intention could have been no more, than to put a stop to the descent of such troops as were intended by the United States, to garrison the posts till they were evacuated: and, as I had no orders respecting the posts, nor men sufficient to garrison them, the order could neither affect me, nor my military escort: and *Thirdly,* that it was impossible that the Governor could have intended that myself and party, should be detained, because it would be in direct violation of the treaty we were preparing to carry into effect. To which the commandant observed, 'that as the waters had risen, one half of the objection at least was done away, and we should meet with no impediment from him.' We then returned to the company.

At four o'clock in the afternoon we took leave of the Commandant, and his officers, whose hospitality and politeness, I shall ever acknowledge with sensibility. As we left the shore, we were again saluted by the artillery of the fort.

Proceeded about two miles down the river and encamped. My mind was occupied the whole evening, in reflecting upon the order of the Governor General, which had been communicated to me by the Commandant at New Madrid. It occurred to me, that if similar orders were given to the other Commandants below, and they should be less liberal and friendly towards the United States, we might be detained several months during a discussion with the Governor General, which from the great distance, would unavoidably be attended with delay, inconvenience and additional expense.

4th, Left the shore about eight o'clock in the morning, and proceeded down the river till sunset.

5th, Got under way before sunrise, but had to put to the shore about ten o'clock in the forenoon in consequence of a violent head wind, which continued till night, accompanied by a heavy fall of rain.

6th, Left the shore early in the morning, but on account of a strong gale of wind, had to make a harbor where we lay till one o'clock in the afternoon, and then continued down the river till after sunset.

7th, Set out at sunrise, put to at the lower end of an island between the first and second bluffs, a few minutes before noon, and took the sun's meridional altitude, then proceeded down the river, and encamped opposite to the third bluff.

8th, Left the shore at sunrise, and continued our journey till a few minutes before twelve o'clock, when we put to and took the sun's meridional altitude, got under way immediately, and arrived at the Chickasaw Bluffs [current site of Memphis], about four o'clock in the afternoon. The Commandant received us politely, but at the same time in a manner, which convinced me that he did not expect us. He appeared somewhat embarrassed, and directed the military escort to encamp on the upper or north side of Wolf river, the fort being below. Mr. Nolan, and myself at his request, accompanied him to his quarters: after being seated, he enquired about an express which had lately gone up the river, with dispatches from the Governor General, to the Commandant at New Madrid. We told him the express had not arrived at that place when we left it, neither had the Commandant received any communication from the Governor General since the preceding November.

The enquiries about the express were very natural, and possibly had no more in view, or to relieve his mind from some anxiety about him, as he had a considerable distance to travel through the wilderness: but the orders from the Governor General, to the Commandant at New Madrid, had produced suspicions in my mind, of which I could not divest myself. I thought it probable that the Governor General had sent on other orders more pointed, and less equivocal, to stop us at that post.[10]

10. Ellicott's suspicions of intentional Spanish reluctance, or ignorance, is a recurrent theme, reinforced no doubt by his own political views of the Spanish and his conversations with Philip Nolan. Such suspicions were confirmed by the constant delays in preparing for the survey, troop movements along the Mississippi by Spanish army and naval officers, and by other evidence. His early suspicions were finally confirmed in his mind when he managed to obtain a copy of a letter (June 16, 1796) written by Governor Gayoso of Natchez to Daniel Clark Sr., a local planter and merchant. The letter read in part: "Spain made a treaty with the Union, but if this Union is dissolved, one of the contracting parties exists no longer and the other is absolved from her engagement. It is more than probable that a separation of several states will take place which will alter the political existence of a power that could influence on the balance of that of others; therefore Spain, being deprived of that assistance which could

9th, Remained at the fort, and though treated in the most polite and hospitable manner, my suspicions were increased from two circumstances. *First,* the Commandant and officers appeared, (or affected,) to be almost unacquainted with the late treaty between the United States and his Catholic Majesty; And, *Secondly,* no preparations either had been, or were making to evacuate that post. Exclusive of these, there appeared an unnecessary suspicion and caution on the part of his Catholic Majesty's officers: two armed gallies were brought into the mouth of Wolf river, between our troops and the fort.

In the course of the day, I informed Mr. Nolan that I strongly suspected something was in agitation respecting the treaty, with which we were unacquainted. He replied, 'keep your suspicions to yourself, by no means let them appear, you may depend upon me, whatever I can discover, you shall know, but the utmost caution will be necessary, both for our success and my own safety.'

10th, Left the Bluffs about eight o'clock in the morning, and put to the shore a few minutes before noon to take the sun's meridional altitude, and then continued our journey till four o'clock in the afternoon, when we encamped at a remarkable place, where the Chickasaws and Chocktaws formerly held their consultations.[11]

11th, Left the shore at sunrise, and proceeded down the river till a few minutes before noon when we stopped to take the sun's meridional altitude, and then proceeded down the river till about three o'clock in the afternoon, when we had to stop on account of a violent gale of wind, which continued till after night.—The thermometer rose that day to 66°. Thunder gust after night, and the musquitoes began to be troublesome.

---

arise from her connexion with the Union, will alter her views. . . . Therefore even if no change should happen in the U.S. the treaty will be reduced to the navigation of this river" (qtd. in Holmes, *Gayoso,* 179). See also Samuel Flagg Bemis, *Jay's Treaty: A Study in Commerce and Diplomacy* (New Haven, CT: Yale University Press, 1923); and, on the political parties and Jay's Treaty, see Noble E. Cunningham Jr., *The Jeffersonian Republicans: The Formation of Party Organization, 1789–1801* (Chapel Hill: University of North Carolina Press, 1967).

11. Ellicott was introduced to the powerful Indian nations (Cherokees, Chickasaws, Choctaws, Creeks, and Seminoles) of the Old Southwest as a novice; however, his own view of Indians, in general, was one of distrust and often disdain. He refers to Indians as "savages" more than once in his writings. The Spanish considered their neighboring Indians as important allies, with whom they negotiated a number of treaties, for their mutual protection against the Anglo-Americans. See Weeks, *Paths to a Middle Ground;* see also Holmes, "The Choctaws in 1795"; and O'Brien, *Choctaws in a Revolutionary Age.*

12th, Set out about nine o'clock in the morning. Cloudy in the forenoon, and rain in the afternoon, attended with thunder and lightning. Stopped at sun down.

13th, Got under way very early, and made no halt till we came to after sun down. Cloudy all day with some rain.

14th, Set off very early, but had to come to at three o'clock in the afternoon, in consequence of a violent head wind.

15th, Left the shore about sunrise. A smart frost last night. Stopped a few minutes before noon and took the sun's meridional altitude, and then got under way, but were brought to about two o'clock in the afternoon by Col. [Carlos] Howard, an Irish gentleman in the service of his Catholic Majesty, who had two armed gallies with him; after detaining us about one hour, we proceeded down the river till sunset and encamped.

16th, Left the shore at sunrise, took the sun's meridional altitude, encamped in the evening. A fine pleasant day.—Thermometer rose to 64°.

17th, Left the shore at sunrise. A cloudy warm morning, and a light shower between seven and eight o'clock, attended with a thick fog which continued about three hours, and separated our boats so completely that they did not all join me till the next day. Took the sun's meridional altitude. At one o'clock in the afternoon put to the shore on account of the violence of the wind. On the shore we found a large alligator, and endeavoured to kill him with our oars and setting [push] poles, but in vain; got under way in about an hour, had a thunder gust in the afternoon. Encamped in the evening. A heavy rain at night attended with hail, sharp lightning and uncommon loud thunder. —Thermometer rose to 65°. Several of our people indisposed.

18th, Left the shore just after sunrise, a fine pleasant day. Took the sun's meridional altitude. The maple (acer negundo) and some other trees began to look green. Encamped in the evening.

19th, Set out early in the morning. Took the sun's meridional altitude on a small island. Near the place of observation, our men killed a large alligator. It appeared to be a very strong, though dull and stupid animal; but the latter might be owing to its just emerging from its torpid state. Arrived at the Walnut Hills [near Vicksburg], where the Spaniards have erected some considerable works.[12] The post [Fort Nogales] is a very important one, and capable of being made very strong. This day was very warm for the season, the thermometer rose to 79°.

12. Walnut Hills and Fort Nogales were also visited and described by Francis Baily in his *Journal,* 146–47.

At this post my suspicions relative to delays being in contemplation by offers of his Catholic Majesty, to prevent the treaty going immediately into effect, were nearly confirmed. The Commandant, though he treated us very civilly when on shore, had us brought to by the discharge of a piece of artillery, which was wholly unnecessary as we were near the landing, and making to it as fast as we could.

After we landed, Mr. Nolan and myself accompanied the Commandant to his quarters, and after taking some refreshment, he enquired my business, with which he appeared to be almost wholly unacquainted, and was only satisfied by my sending to the boat for my papers, and producing an authenticated copy of the treaty in his own language! That the Commandant could be unacquainted with a transaction so notorious, and stationed at a post in the vicinity of a settlement, in which Governor Gayoso resided, appeared very extraordinary, and could only be considered as the effect of affectation.

20th, The commandant very politely conducted us to their different works. At noon I took the sun's meridional altitude at the curtain of the lower battery; after which we dined with the Commandant and his officers, and then proceeded down the river, and encamped about sun down.

The Walnut Hills are so called from the tree of that name (juglans nigra,) with which they formerly abounded. The situation is handsome and commanding, but said to be unhealthy: small beautiful stream of water, falls from the hills into the Mississippi, which the Commandant informed us was, notwithstanding its favourable appearance, of a very bad quality. A number of peach trees which had been planted on the hills were now in full bloom.

21st, Set out at five o'clock in the morning, but were driven on shore about nine o'clock in the forenoon by a violent gale of wind, which continued till some time in the night. Took the sun's meridional altitude.—Thermometer 74°, almost the whole day, and musquitoes from that degree of heat had become very troublesome.

22nd, Got under way about four o'clock in the morning. A few minutes after we had left the shore, I received the following letter from Governor Gayoso. [Gayoso spoke fluent English.]

Sir,

Some gentlemen that left you at the mouth of the Ohio, have informed me of your approaching arrival here, and that to attend you on your commission you bring a military guard and some woodsmen.

It is with pleasure that I propose myself the satisfaction of seeing you here and to make your acquaintance.

Though I do not conceive that the least difficulty will arise respecting the execution of the part of the treaty in which you are an acting person, yet as we are not prepared to evacuate the posts immediately for the want of the vessels that I expect will arrive soon, I find it indispensable to request you to leave the troops about the mouth of Bayou Piere, where they may be provided with all their necessaries, which you can regulate on your arrival here. By this means every unforeseen misunderstanding will be prevented between his Majesty's troops and those of the United States, besides it is necessary to make some arrangements previously to the arrival of the troops, on which subject I shall have the honour of entertaining you when we meet.

I embrace this opportunity to assure you of the satisfaction I feel in being appointed to act in concert with you, though your first interview is to be with the General in Chief of this Province.

I have the honour to be with the highest consideration, Sir, your most humble servant,
(Signed) Manuel Gayoso de Lemos.

Honorable A. Ellicott.

The foregoing letter was sent by an express to the Walnut Hills by land, but we having left that place before he arrived, the Commandant manned a light canoe, and sent him after us.

CHAPTER III.

Containing official correspondence with the officers of his Catholic Majesty, connected by a detail of the circumstances which produced it—interview with a mysterious character from Philadelphia—continuation of the official correspondence and detail, with the incident which produced the general commotion among the inhabitants—the election of a committee, and termination of the commotion by a compromise between the Governor and inhabitants.

The correspondence with the officers of his Catholic Majesty and other persons, and the incidents relative to the political state of the country contained in the ensuing part of this work are substantially the same, as detailed to the Secretary of State, and partly copied from my original communications: some things are added which it was not thought advisable to trust to a hazardous conveyance; but no opinion is altered, and no new colouring is given to meet either the changes of sentiment, of men, or of measures.

Immediately after we arrived at the town of Natchez, the following note was sent to Governor Gayoso by Mr. Nolan.

> Landing at the Town of Natchez, Feb. 24th, 1797.
>
> Sir,
>
> It is with pleasure that I announce to you, my arrival as commissioner on behalf of the United States, for carrying into effect the third article of the treaty lately concluded between the said United States, and his Catholic Majesty, I wish to be informed, when it will be convenient for your Excellency to receive my credentials.

I am, sir, with due respect,
Your humble servant,
Andrew Ellicott.

His Excellency Manuel Gayoso de Lemos.

In the evening the following answer was received by the hands of his secretary Mr. [Joseph] Vidal.

Natchez, 24th Feb. 1797.

Sir,

By your favour of this day, delivered to me by Mr. Nolan, I learn with pleasure your arrival at this post, in the character of commissionary in behalf of the United States, to ascertain the boundaries between the Territory of his most Catholic Majesty, and that of the said United States.

I have the honour to be, with the highest respect, Sir, your most humble servant,
Manuel Gayoso de Lemos.

Honourable Andrew Ellicott.

The Governor in his reply, having made no answer to that part of my note respecting the reception of my credentials, the time when they could be presented of consequence remained uncertain: several verbal messages passed between us, before this point was adjusted. The Governor urged that we had come upon him by surprise, that it would have been proper to have remained at some distance above the town, till he was officially informed of our approach: but it being now too late to go through this ceremony, the Governor at length consented that we should meet at the government house in the afternoon of the 25th. The meeting took place accordingly, and immediately after my credentials were produced, the Governor was pressed to name a day on which our operations should commence. On this subject some discussion took place, which ended with the Governor's naming the 19th of March following.

On the 27th of February, the following letter was written to the Governor General the Baron de Carondelet, at New Orleans, who was named by his Catholic Majesty as principal commissioner.

Natchez, Feb. 27th, 1797.

Sir,

It is with pleasure, that I embrace this opportunity of informing you of my arrival at this place, as commissioner on behalf of the United States, for ascertaining the boundaries between the territories of his Catholic Majesty, and those of the said United States.

The polite manner in which I have been received at the posts on the Mississippi, now in the possession of his Catholic Majesty, demands my thanks and gratitude, and I am in hopes that a similar conduct will be observed on our part.

I have the honour to be, with great esteem,
Your Excellency's humble servant,
Andrew Ellicott.

His Excellency the Governor General
The Baron de Carondelet.

On the same day that the letter was written to the Baron de Carondelet, we encamped on the top of a hill, at the upper end of the town about one quarter of a mile from the fort [Spanish Fort Panmure de Natchez], and on the 29th hoisted the flag of the United States.[13] In about two hours after the flag was hoisted, a message was received from the Governor directing it to be taken down! This request met with a positive refusal, and the flag wore out upon the staff. We were several times notified that parties were forming to cut it down, but the attempt was never made; it would have been resisted by force. On the day the flag was hoisted, a course of astronomical observations was begun, the observations will be found in the Appendix.

Before we encamped, the following intelligence was communicated to me through confidential channels. *First,* that in September previous to my arrival in that country, the Baron de Carondelet in a private conversation, declared the treaty would not be carried into effect, that he as principal commissioner

13. Governor Gayoso regarded the flying of the American flag on Spanish territory as a diplomatic affront, but reflective of poor timing more than anything else since Spain had not yet evacuated the post. He continued to try to make additional concessions to keep the peace, especially with the Indians in the district. See the descriptions of these tense days provided by Holmes, *Gayoso,* 183–86; Cox, *The West Florida Controversy,* 40–45; and Whitaker, *The Mississippi Question,* 52–55.

should evade, or delay, from one pretense or other, the commencement of the operations. *Secondly,* that a letter had been written bearing date June 16th, 1796, by governor Gayoso, to a confidential friend [Daniel Clark Sr.], stating that the treaty was not intended to be carried into effect, and that delay on their part would reduce it to a dead letter. (The original letter has been in my hands.) And, *Thirdly,* that the country either was, or would be ceded to the republic of France. [Such negotiations between Spain and France had been ongoing since 1795.] This intelligence was kept a profound secret for two reasons: *first,* because its being known might have produced suspicions injurious to individuals, and *secondly,* that we might be able gradually to effect our object, and secure to the United States a country very important both from its situation and value of its commerce, before any direct explanations should take place. The dispositions of the inhabitants were sounded, and a large majority appeared in favour of becoming citizens of the United States.

The dispute between the United States and the republic of France, presented serious difficulties, owing to the alliance between the latter power and his Catholic Majesty.[14] It did not appear improbable, that the common cause would be made between those nations against the United States, and Great Britain, who had lately entered into a treaty [Jay's Treaty], which was offensive to the republic of France, and to say the best of it, was very unpopular in the United States.

Whatever my prejudices might be in favour of the principles of the French Revolution, and of that nation, for the part it took in our arduous struggle with Great Britain for the liberty we now enjoy, I considered it my duty, as a citizen of the United States, not only to retain the post we then occupied, but to extend our limits if hostilities should commence.[15] This is not meant

14. The period from 1797 to 1800 is known as the Quasi-War between the United States and France. On March 2, 1797, the French government authorized the seizure of neutral vessels, an act aimed at American shippers. In response, the Federalist-dominated Congress followed by declaring the 1778 treaty with France null and void in July 1798. The Quasi-War between France and the United States ended with the Convention of September 30, 1800. But the next day, October 1, 1800, Spain secretly ceded Louisiana back to France, which took the "Mississippi Question" to an entirely new level. Unable to halt the advances of the Americans into the Mississippi Valley, the gateway to Mexico, Prime Minister Godoy had a better idea—let the French do it! See Alexander DeConde, *The Quasi-War—The Politics and Diplomacy of the Undeclared War with France, 1797-1801* (New York: Scribner's, 1966).

15. Ellicott here is at odds with his own Federalist Party, which supported Great Britain. Secretary of State Timothy Pickering especially was adamantly anti-French, and had advocated

as an apology for my conduct, but a declaration of my sentiments. My commissary Mr. Anderson was directed to procure all the ammunition he might find among our friends, but to do it in as private a manner as possible. My party then consisted of about 30 persons, (exclusive of the escort which yet remained at the Bayou Piere [Pierre]), well acquainted with the woods, and generally armed with rifles.[16]

Notwithstanding these arrangements, peace was considered the true interest of the United States, and positive orders were several times given to treat the Governor, and other officers of his Catholic Majesty, with all the respect due to their stations, and to which they were particularly entitled from us, for the attention and hospitality we had experienced from them: and that if hostilities should be commenced, we might not be the aggressors.

We had been but a few days at Natchez, before the Indians became very insolent, insulted a number of our men, walked about the camp with drawn knives, and one night we were informed that they intended attacking us, and they actually came part of the way from their camp towards ours, but whether for the purpose suggested I am not certain. In consequence of these repeated insults, and to be able to make some stand if attacked, the following letter was written to Governor Gayoso.[17]

---

declaring war during the Quasi-War era. Federalists generally regarded the French Revolution as the rule of the mob; Spain's treaty of peace at Basel in 1795 with France was regarded as symptomatic of its inherent weakness, with Spain, a "feeble" nation, now merely a tool of the French.

16. In addition to the thirty men for his workforce, Ellicott noted that his military escort included another twenty-five soldiers under the command of Lt. "Crazy" Percy Smith Pope, and under the orders of General "Mad" Anthony Wayne. Whitaker states that, according to Carondelet, Ellicott had with him "thirty laborers and twenty-five soldiers" (*The Mississippi Question,* 280). By comparison, Governor Gayoso had about fifty soldiers at the fort in Natchez. However, an assessment of the military situation in Natchez at this time should take into account the pro-American attitude of the inhabitants, who could quickly become militia units with which to supplement any American regular troops. See Holmes, *Gayoso,* 184–89; and, for troop locations, see Ellicott's map (map 3).

17. Ellicott regarded the Spanish as stirring up the Indians. Alliances with the Indian nations formed a cornerstone of Spanish policy in the Mississippi Valley. For the Indians, treaties with the Spanish permitted the latter to construct defensive forts, not settlements that took their land, and from which the Indians, too, could be protected from the Americans. In the treaty negotiations, the Spanish always gave presents, usually guns and other necessities. The Indian nations (Creeks, Cherokees, Choctaws, Chickasaws, and Seminoles) of the Old Southwest consisted of warriors who could live off the land, endure hardships and long marches, and, when properly led, prove very effective military units. They were among the finest light infantry in

Natchez, March 11th, 1797.

Dear Sir,

The conduct of the Indians yesterday, and last night, owing principally to their constant state of intoxication, renders it absolutely necessary in my opinion, to have recourse to my military escort for protection.

The discipline of our army is such, that you may rest assured none of the inconveniencies mentioned in your first communication to me, are to be apprehended from the escort being stationed at this place: And as the attendance of the guard, forms a part of the treaty now carrying into effect between the United States and his Catholic Majesty, and which I am authorized to declare will be observed by the nation I have the honor to serve, with good faith and punctuality. From these considerations, I must request the favour of you, to withdraw your objections against my escort's joining me at this place as soon as possible.

I am, Sir, with great esteem and respect,
Your Excellency's friend,
And humble servant,
Andrew Ellicott.

His Excellency Manuel
Gayoso de Lemos.

Immediately after sending the foregoing letter to Governor Gayoso, the following was received from Governor General the Baron de Carondelet.

Translation.
New Orleans, 4th March 1797.

Sir,

I received with much satisfaction your favour of the 27th of February last, in which you are pleased to communicate to me your arrival at

---

the world. Governor-General Don Bernardo de Galvez had made excellent use of his Indians units, along with Spanish regulars, Louisiana militia, and mercenaries during his campaigns from 1779 to 1781 in both Louisiana and Florida against the British. See William S. Coker and Robert R. Rea, eds., *Anglo-Spanish Confrontation on the Gulf Coast during the American Revolution* (Pensacola: Gulf Coast History and Humanities Conference, 1982). Governor Gayoso participated in several Indian treaties on behalf of Spain (see Holmes, *Gayoso,* 141–61, 235–37).

Natchez, in the character of commissioner on the part of the United States of America, for settling the boundary line between the territory of his Catholic Majesty, and the said United States. I am equally well pleased, with the testimony you have given of the civilities and attentions which were shewn to you from the commanders of the different posts; which has been conformable to the intentions of the Sovereign, to my orders, and to the general principles of the nation, and I doubt not but that on every occasion Spaniards would receive a like attention from the citizens of the United States.

May God preserve you, etc.
Baron de Carondelet.

Mr. Andrew Ellicott.

It may be observed, that the Baron de Carondelet wholly omitted saying any thing about commencing our operations, this omission was not expected, though I was well aware that delay would be attempted.

Very soon after the letter was received from the Baron, Governor Gayoso called at my tent, and informed me, that the Baron in consequence of some pressing concerns below, had declined to attend, and that the whole business had devolved on him, with which I expressed my entire satisfaction, and particularly as we were both on the ground, we might immediately proceed to make our arrangements. He assured me, that no time should be lost, although it would be impossible for him to be ready by the 19th [of March]. He added further, that though the Baron could not attend on account of other pressing business, he was nevertheless very desirous of having an interview with me at New Orleans, and for that purpose had given orders to have a galley fitted up for my accommodation to New Orleans, and from thence, wherever my curiosity might direct. In reply, the Governor was informed, that the proposal would be taken into consideration, and answer given the next day. The day following, a number of gentlemen (in the Spanish interest,) waited upon me with assurances, that the Baron was anxious to see me at New Orleans, that he had written to them upon the subject, and that I might depend on every attention in his power. By this time my mind was made up. The invitation appeared calculated to produce delay, divide my small party, and leave no rallying point for such of the inhabitants of the district, as were in our favour, if the commotions and intrigues in Europe should produce hostilities between the United States, and his Catholic Majesty. From these considerations the invitation was not accepted. Having an opportunity the same day of writing to New Orleans, the following letter was written to the Baron de Carondelet.

Natchez, March 12th 1797.

Sir,

Your favour of the 2nd of this month has been duly received, but contrary to my expectation, does not contain any information respecting the time you will be ready to proceed to the ascertaining of the boundaries between the United States and his Catholic Majesty. This appears to me the more extraordinary, as his Excellency Governor Gayoso informed me in his first communication, that your Excellency was the person with whom I must have the first interview relative to the arrangements, and manner of carrying into effect that part of the late treaty, so far as it respects the boundaries between the nations which we represent.

Although my detention here at a great expense to the United States, gives me much uneasiness, my concern would be greatly increased, could I for a moment suppose, that any impediment would be thrown in the way, to prevent a speedy and full execution of a treaty, in which both nations from their local situations are so deeply, and mutually interested: that these interests, and a reciprocity of good offices, may produce a perpetual friendship between the nations we have the honour to serve, is the sincere wish of

Your Excellency's friend, and
Humble servant,
Andrew Ellicott.

His Excellency the Baron
de Carondelet.

In the afternoon of the same day, the following letter was received from Governor Gayoso.

Natchez, 12th March, 1797.

My Dear Sir,

This morning I had the pleasure to receive your amicable communication dated of yesterday. I give you my sincere thanks for having established this form of intercourse, as it will make our business more easy, and indeed it is more conformable to the sincere friendship that we have contracted.

In answer to your said letter, I will remark, that such conduct of the Indians is not customary here, I foresaw that it would happen, from the moment you shewed a desire of having your colours flying before all the transaction were terminated; knowing the Indians as well as I do this was the reason of the objection I offered the moment I saw it hoisted, for otherwise I know very well, that it is frequently used by the representatives of any nation in a foreign country, it is even done in Spain by foreign consuls. I am sorry that you should have experienced any inconvenience from this particular circumstance, and that urged by such effect, to wish to have by you your escort. I have not the least objection that it should be called from its actual station, but as it is my duty, and that I am answerable for the tranquility of the country that is entrusted to my charge, I must propose to you a method that will answer every good and satisfactory purpose. Had you not been unluckily stopped on your voyage to this country, you would have had immediately the General [Carondelet] of the province here to begin the operation, of demarking the divisor line between the territories of his Catholic Majesty, and those of the United States of America, he had every necessary to attend to the business, but since the time that he had a right to expect the commissioner of the United State, the war with England has taken place, and his cares thereby increased, yet he expected to have it in his power, to come to meet you at Daniel Clark's, Esquire, which place is near the point of the 31°, but he has found it impossible, as it would oblige him to make too long an absence from New Orleans, therefore it is myself that will have the honour to accompany you on that important commission on behalf of his Catholic Majesty. This is the moment that I am in want of every individual thing, both for my person, and for the attendants of the commission, though the geometer, and other officers that are to be employed, are already on their way from New

Orleans, and will stop at Clarksville, where I shall go myself as soon as my equipage arrives from the capital, but this will inevitably take some time, therefore the plan that I wish to arrange with you, will be to make Lofftus's Cliffs [Fort Adams] our point of re-union, this place is a short distance from Clarksville, and it is very healthy situation, there I will send everything concerning the Spanish commission and that will be the most convenient place to establish for a while your headquarters under your military escort. By adopting this measure, you will have all your people together, and the most distant disagreeable occurrence avoided; as I am positively confident that some would happen by the conjunction here as you propose. It is true that by the treaty an escort is supposed, and even recommended to each commission, but it is to be on the line, and not at a distance from it, where it would interfere with other business, therefore I feel sensibly hurt that it is out of my power to consent in the landing of the troops at this place, though I have not the least objection on their going directly to Lofftus's Cliffs.

I have given the most positive orders to prevent the Indians getting liquor, and to their interpreter I have given the strictest charge to be always in sight, and to-morrow I expect that they will remove to some distance from hence.

I have the honour to be, with the highest
Respect and esteem,
Your most affectionate
Friend and humble servant,
Manuel Gayoso de Lemos.

The Honourable
Andrew Ellicott.

By this letter, it will be seen that my request relative to the military escort had not the desired effect: but in the meantime, an express had been privately sent to the commanding officer of the escort, with directions to proceed down to the town of Natchez.

Shortly after the Governor's letter was received, we had an interview. He immediately undertook to prove the propriety and necessity of our going down the river to Clarksville, where myself and party, might be altogether; and closed his reasoning by declaring, that if the escort did land at the town of

Natchez, he should consider it an insult offered to the king his master. He was told, that few observations would be made at present, as his letter should be replied to immediately, but that there was one circumstance which appeared to require some explanation; that was, the desire constantly manifested to draw us from the town of Natchez, to some other place, less convenient, and more out of the way of information: but as the town of Natchez was designated by the treaty as the place of meeting for the commissioners, any propositions to draw us from thence would be rejected without ceremony. To which the Governor replied, 'you have either mistaken my meaning, or I have expressed myself very badly, I do not want you to leave this place, on the contrary, I am desirous you would take up your residence at my house, where you would be much more comfortable than in a tent.' He was told in reply, that the tent was more agreeable than a palace, and in which that independence, so characteristic of the country I had the honour to serve, could be indulged and gratified.

On this subject the Governor and myself no doubt perfectly understood each other: he wanted us from the town, and in some place of small importance, and distant from the principal settlement, the fidelity of whose inhabitants he suspected; or at his own house, where our intercourse with them would be under his own eye, and where our plans could not easily be carried on, and digested without a discovery: and for the express purpose of making use of the inhabitants to carry the treaty into effect, or secure the country by force, if such a measure should become necessary, was our real motive for continuing at that place.

On the 13th at noon, the following was sent to Governor Gayoso, as a reply to his letter of the preceding day.

Natchez, March 13th, 1797.

Dear Sir,

Your favour of yesterday would have been answered sooner, had not the storm last night prevented me from writing in my tent.

Your letter, as well as other circumstances that have come to my knowledge, contain fresh proofs of your desire to promote good order and harmony in this part of the country. But sir, I cannot suppose that any inconvenience could possibly arise, or the peace of the settlement be disturbed, by the arrival and landing of the escort which I left at Bayou

Piere. If I did suppose the contrary, I trust that I should be one of the last persons to propose the measure. In my opinion the escort which accompanied me to Bayou Piere, is as much bound to observe good order in this country, as the troops of his Catholic Majesty. This is not an opinion of the day, it has uniformly been mine ever since I left the seat of our government, in consequence of which immediately upon my entering the Mississippi, I issued standing orders, that when any of our party, the military included, should be in any place the jurisdiction was exercised by his Catholic Majesty, the laws and usages of that government should be observed, and submitted to, in the most pointed manner.

As I hope that mere punctilios will never interrupt our friendship, and as the conduct of the Indians has become less exceptionable, since the night before last; I am not so anxious for the escort to be stationed at my present encampment. I would therefore, to prevent any misunderstanding or disturbance, propose that the officer who commands the escort, be directed to proceed down the river to Bacon's landing, from whence he may come to this place, and procure such necessaries as he may be in need of for the season.

As this is the place designated by the treaty for our meeting and making the arrangements for carrying on the business, I conceive there would be an impropriety in my leaving it, until your Excellency is ready to join in fixing the first point of latitude.

I have the honour to be,
With great esteem,
Your humble servant,
Andrew Ellicott.

His Excellency Manuel
Gayoso de Lemos.

I now found myself involved in a dilemma, the troops were by that time (by my order) on their way to Natchez, and that contrary to the orders of the Governor: being unacquainted with intrigue and address, it was determined to support openly the descent, and landing of the escort, and which might have been done upon good ground, and fair argument, if no other considerations had been involved in the Governor's objections; but the discussion was rendered unnecessary by the following letter from the Governor.

Natchez, 14th March, 1797.

My Dear Sir,

I do myself the pleasure, to acknowledge the reception of your favour of yesterday, and am very happy to find that our sentiments uniformly agree in every thing that can combine the mutual interests of our nations, and I pledge to you my honour and friendship, that every step in my conduct shall be guided by this principle, impressed in me by my duty, and by the very particular attachment that I have for you.

I have the honour to be, with the greatest esteem and affection,
Dear Sir,
Your most humble and obedient servant,
Manuel Gayoso de Lemos.

Honourable Andrew Ellicott.

The Governor's letter was handed to me by his aid Captain Stephen Minor, who was asked, whether the Governor meant by his polite communication that the escort might be stationed at Bacon's landing, or to remain at the Bayou Piere, the latter now being impossible, as I hourly expected its arrival. The captain answered, that 'the Governor acquiesced in its being stationed at Bacon's landing.'

On the 15th, in the evening, the escort arrived, and on the afternoon of the 16th it proceeded down to its station.

A few days after the landing of the escort, the following verbal message was received from the Governor, by his aid captain Minor. 'Sir, his Excellency has been informed, that the officer commanding your escort, has taken up a number of men as deserters from your army, some of whom are detained contrary to their inclinations, he therefore requests that they may all be immediately discharged.' Captain Minor was desired in return, to inform the Governor, that a compliance with his request would require some previous consideration, that it was a subject which might be of importance to both nations, in the prosecution of our business, and therefore the line of conduct we intended to pursue, ought to be well designated to prevent disagreeable interferences: that the next day, the Governor should have my answer. And, accordingly in the presence of William Dunbar, Esq. (a gentleman whose talents, extensive information and scientific acquirements, would give him a distinguished rank in any place or in any country,) Captain Minor was desired to inform Governor Gayoso, that his request relative to the deserters had been taken into consideration, and that my conduct should be regulated by the following view of the subject, viz. *First,* That all the deserters from the army of the United States, who came into that country since the period appointed by the late treaty, for the evacuation of the posts held by his Catholic Majesty within the limits of the United States, should be considered as wholly without the protection of his Catholic Majesty, and liable to be taken, and detained whenever they fell in our way. *Secondly,* such deserters, as they came into that country before the time stipulated for the evacuation of the posts, and had taken protection of the Spanish government, should, as their case was somewhat doubtful, remain for the present unmolested. *Thirdly,* that all persons against whom there were standing proclamations by the executive of the United States, if found north of a line to drawn east from the Mississippi, and thirty-nine miles south of the town of Natchez, should be apprehended if they fell in our way. Here this subject appeared to rest.

About the time my escort arrived, the principal part of the artillery was taken out of the fort [Fort Panmure], and carried to the landing, and there was every appearance of a speedy evacuation; but on the 22nd, great industry was used in taking it back, and the cannon were immediately remounted. This gave great alarm to the inhabitants of the district, who generally manifested a desire of being declared citizens of the United States, and at once to renounce the jurisdiction of Spain. In order to quiet the minds of the inhabitants, and be able to give them some reason for the Governor's conduct, which they now began to consider as hostile towards the United States, the following letter was written to him on that subject.

Natchez, March 23rd, 1797.

Dear Sir,

The remounting of the cannon at this place, at the very time when our troops are daily expected down to take possession of it, the insolent treatment which the citizens of the United States have lately received at the Walnut Hills, and the delay of the business (on your part) which brought me into this country, concur in giving me reason to suppose, that the treaty will not be observed with the same good faith and punctuality, by the subjects of his Catholic Majesty, as it will by the citizens of the

United States. I hope your Excellency will give such an explanation of the above, as to remove my doubts and apprehensions, which I am afraid have been too justly excited.[18]

I have the honour to be, with great
Esteem and respect, your friend and
Humble servant,
Andrew Ellicott.

His Excellency Manuel
Gayoso de Lemos.

The foregoing letter was followed by the following Note.

Mr. Ellicott presents his compliments to his friend Governor Gayoso, and wishes to be informed, whether it is true or not, that all the works at the Chickasaw bluffs have been either demolished, or carried to the opposite side of the river, and that every exertion is now making at the Walnut Hills to put that post in a complete state of defence. This representation has this moment been made to Mr. Ellicott.
March 23rd, 1797.

This letter and note of the 23rd, produced the following reply.

Natchez, 23 March, 1797.

My Dear Sir,

I have just received your communication of this day, by which I am sorry to find the construction you put on the storing of the ammunition that came from the Walnut Hills in this fort, I have no other place to put them in, for it would be imprudent to leave them exposed in an insecure place, in a time that the Indians might take advantage of us, if they found that in the present circumstances we acted without the necessary precautions. At the same time, that you see me conducting ammunition to

18. Governor Gayoso did not formally object to Ellicott's military escort under Lieutenant Pope from coming down the Mississippi River to Natchez in March, but Ellicott by this time was suspicious of any Spanish actions, and their remounting of the cannon at Fort Panmure de Natchez seemed to reinforce his views.

the fort, you will likewise see as many got out of it for the Arkansas, to reinforce that post, which now will be exposed to the incursions of the Osage Indians, who in the last season pillaged the white hunters of that country.

I am entirely unacquainted of any ill treatment, that the citizens of the United States should have received at the Walnut Hills; if you mean the execution of the orders of the General in Chief [Carondelet] of this province, to demolish that post, it was in consequence of our treaty with the Indians, that they might have no just reason to complain of our conduct, but since I have been informed of their unsettled disposition, I have sent counter orders to suspend every thing that might injure the actual estate of those fortifications, and in such circumstances shall not move any thing else until the arrival of the American troops that are daily expected.

The unavoidable detention that has been experienced in beginning the line, you know the reasons, but they shall soon be removed, as Lieutenant Colonel [Gilberto] Guillemard is far on his way up, and at his arrival this important business shall be begun.

I do assure you, that there is nothing that can prevent the religious compliance of the treaty, though I might observe, that the conduct of some persons that seem to affect an immediate interest for the United States is such, as to occupy my attention. I request that you will be so kind as to take such measures, as to suppress untimely expressions that can only tend to disturb the tranquility of the public, of which I am solely answerable for the present. As I was finishing this, your secretary, Mr. Gillespie, brought me your note, enquiring if the works at the bluffs had been destroyed or removed to the other [Spanish] side of the river.

What I have already said concerning our treaty with the Indians, I suppose, has guided the Governor General of this province to take that step. I really do not know whether they are destroyed or not. I give you my word, that I did not know what was to be done there, and it is only by Baron [Philippe Enrique Neri de] Bostrop that I learn, that post would soon be evacuated; but as this is a thing that only regards the General [Carondelet] of the province, I cannot account for it, nor can I say a word more on the subject, as all the orders proceed from him, that post being entirely out of my jurisdiction.

I am with the highest esteem and respect,
Dear Sir,
Your most humble and obedient servant,
Manuel Gayoso de Lemos.

Honorable Andrew
Ellicott.

Governor Gayoso in this letter showed less caution, address and judgment, than in any other one he wrote during the whole discussion. He begins by stating, that the carrying back and remounting the cannon, was no more than storing ammunition which had been brought down from the Walnut Hills: when the reverse was manifest to every person in the country, and the whole transaction within view of my tent. No military stores whatever had been brought down from the Walnut Hills, on the contrary, they were making that post more tenable, and the ammunition which he pretended to be sending to the Arkansas, was actually intended for, and sent to the Walnut Hills: and again, the absurdity of sending military stores down the river from the Walnut Hills to Natchez, and from there back to the Arkansas was too glaring to merit any serious animadversion. Their fears respecting the Indians were merely affectation, the Indians at that time were decidedly in their favour, and no measures had been taken by our administration to secure the friendship of Chocktaws, (a brave and numerous nation,) and through whose country we had to pass.

Notwithstanding the inconsistency of this letter of the Governor's, I felt inclined to take it up in a serious point of view, but gave way to the opinion of my particular friends, or little council, which consisted of a few of the best informed, intelligent and independent gentlemen of the district, by whom it was thought best for the present to make a short reply, and affect to have no doubt of the truth of the Governor's statement: in consequence of which the following was sent to him the next day.

Natchez, March 24th, 1797.

Dear Sir,

It is with pleasure I acknowledge the receipt of your Excellency's very satisfactory letter of yesterday. You may rest satisfied, that I have, and shall continue uniformly to discountenance every measure, and the propagation of any opinion, which may have a tendency to disturb the good order and harmony of this settlement.

I shall close this with requesting, that the Commandant at the Walnut Hills, be directed to treat the citizens of the United States with politeness when they stop at that post, as a contrary conduct may be attended with disagreeable consequences on a river, which both nations have an equal right to navigate.

I am, sir, with great respect and
Esteem, your friend and
Humble servant,
Andrew Ellicott.

His Excellency Manuel
Gayoso de Lemos.

From this reply, the Governor supposed that he had allayed my suspicions, and that he might now venture one step further, and make use of me as an instrument in stopping some of our troops that were said to be on their way down the river under the command of Lieutenant Pope. For this purpose, he sent to me by his aid Captain Minor, an open letter to Lieutenant Pope, informing him, that for sundry reasons it would be proper, and conduce to the harmony of the two nations for himself and the detachment under his command to remain at or near the place where that letter should meet him, until the posts were evacuated, and as every preparation was making for that purpose, the delay would be but for a few days, when he would be happy to see him at Natchez. This proposal to Lieutenant Pope he wished me to second, as may be seen in the following letter.

Natchez, 25th March, 1797.

My Dear Sir,

By every report, you are acquainted with the confirmation of every thing I have told you concerning our business, you know that Lieutenant Colonel [Gilberto] Guillimard will be here soon, and that immediately we shall proceed to the running of the [boundary] line. But as nothing than friendly arrangements are to guide our conduct, it is necessary to avoid every shadow of compulsion. By the contents of my letter to Mr. Pope, you will see my reasons, therefore request that you will join a couple of lines to avoid any more writing.

I am hurried by many people that have business, this being court day, though I have tried to disembarrass myself, but cannot wait upon you.

I am with the highest esteem and
Respect, my dear sir,
Your most humble servant,
Manuel Gayoso de Lemos.

Honorable Andrew
Ellicott.

After reading the Governor's letter, Captain Minor was informed, that it was impossible for me to join with the Governor in his request to Mr. Pope, as it was well known to me, that instead of evacuating the posts, they were making them more defensible; however I would write to Mr. Pope, and requested him the captain, as he was to be the bearer of the Governor's letter, to take charge of mine also, to which he consented: it was in the following words.

Natchez, March 25th, 1797.

Sir,

This will be handed to you by Captain Minor, a friend of mine, and an officer in the service of his Catholic Majesty, your polite attention to him will be considered as a favour conferred on me.

By order of Governor Gayoso, his letter to you of this date has been shewn [shown] to me; his request for you, and the troops under your command, to remain an indefinite length of time above this place, appears to me a very extraordinary one: as sufficient time has already been given by the United States, for evacuation of all the posts on the east side of the Mississippi above the 31st degree of north latitude, and from the circumstance of the cannon belonging to this place, after being taken to the landing apparently for transportation, being taken back and remounted, I cannot pretend to say that an evacuation is really intended in any reasonable length of time.

From this, and some other considerations, I should conclude that the sooner you are here the better. However, as I have no control over the destination of the troops of the United States, except my own escort, I shall take it for granted that your instructions are sufficiently pointed to direct your conduct.

Please to accept of my sincere wishes for the safe and speedy arrival of yourself and troops to this place.

I am Sir,
Your friend and humble servant,
Andrew Ellicott.

Lieutenant Pope.

The letters were taken up to the Walnut Hills, and safely delivered by the Commandant to Lieutenant Pope on his arrival, which was eight or nine days after Captain Minor had returned from that place.

A few days before Captain Minor went up to the Walnut Hills, a gentleman arrived at Natchez, who informed me that he had a communication to make which was for myself only. It was immediately arranged, that the communication should be made in my tent the next day at nine o'clock in the forenoon. At the time appointed he attended, and after common ceremonies were over, he [John Chisholm of Tennessee?] began a panegyric upon the late Mr. William Blunt [Blount] of the state of Tennessee, observing that for his knowledge of men, and dexterity at intrigue, he was perhaps unrivalled in the United States, and that the secretary of war [James McHenry, Federalist from Maryland, who served as secretary of war from 1796 to 1800] had consulted him (the stranger) upon the propriety of sending Mr. Blunt to France, as minister plenipotentiary to adjust our differences with that nation. Our opinions respecting Mr. Blunt not coinciding, no communication was made, unless his informing me that the secretary of war was aware of the difficulties we should meet with in carrying the treaty into effect before he left the city of Philadelphia, (which was near the latter end of the preceding December) can be called a communication.

If those difficulties were known to the administration at that early period, it was certainly very improper to leave us in the dark to combat them, without having any line of conduct marked out for us to pursue. It left it in the power of the administration to avow or disavow our proceedings, as they were more or less successful. This gentleman, after remaining a few weeks at Natchez, and in its vicinity, and spending some time with Mr. Anthony Hutchins, who was then a Major on the British military establishment, and with a Mr. Rapleja [Englishman, Dr. Nicholas Romayne] a refugee from the state of New York, and likewise on the British establishment, he went to Mobile, and Pensacola, where he resided at the house of Panton, Laslie [Leslie], & Comp. [English trading company with a Spanish monopoly among the Indians] and their connexions until the explosion of Mr. Blunt's plans [1797].[19] Panton, Laslie, & Comp. are equally known to both Spanish and British governments, as British subjects. For a considerable annuity which they pay to the King of Spain, they have the exclusive privilege of trading with the southern Indians east of the Mississippi, from which circumstance they have acquired a great influence over them. Of this distinguished house, I shall have some occasion to speak hereafter. The gentleman on whose account this digression has been made, and whose mysterious conduct embarrassed me exceedingly, was paid for his services, whatever they were, by the public.

On the 29th of March the Governor issued the following Proclamation.

19. William Blount (1749–1800) was from North Carolina and served in its delegation to the Constitutional Convention in 1787. He moved to Tennessee, where he acquired large land allotments and was deeply in debt for land speculation there. His work in opening new lands led to his appointment as governor of the new Southwest Territory (1790–96). When Tennessee became a state in 1796, he was elected to the U.S. Senate. His western land speculations, however, put him deeply in debt, and when news of the Franco-Spanish Treaty of Basel (1795) became known, he assumed that Spain would soon return its Louisiana province to France. If this were to happen and French were to regain control of the Mississippi Valley and New Orleans, Blount's speculations in western lands would become liabilities. His only option, as he saw it, was to negotiate a deal with Great Britain that would grant him and his associates—John Chisholm, John Brown, Benjamin Sebastian, John Fowler, Harry Innes, and Caleb Wallace—certain concessions in the West in return for their assistance against the French. Blount's plans were discovered, and Secretary of State Pickering and President John Adams sent the incriminating evidence to the United States Senate in 1797, which began impeachment proceedings against him. It was public knowledge of these events that reinforced Spanish officials' beliefs that the Americans and British were in league with one another, and that an attack on Louisiana was imminent. Governor-General Carondelet ordered the assembling of 2,500 troops at Baton Rouge as a step toward halting any Anglo-American invasion. In 1799, the impeachment proceedings against Blount were dismissed; he died in Tennessee in 1800.

Don Manuel Gayoso de Lemos, Brigadier of the Royal Armies, Governor Military and Political of the Natchez, and its dependencies, etc. etc.etc.

WHEREAS the political situation of this country, offers a large field to busy, and malignant minds, to agitate and disturb the tranquility of its inhabitants, it is therefore my duty, and in the continuation of that vigilance which I have constantly exerted, not only to promote the happiness of every individual of this government; but likewise to support their interest, and secure their tranquility, that I now step forth to warn the public against being led by their innocent credulity, into any measure that may be productive of ill consequences, and frustrate all the advantages that they have a right to expect, and that by the present I assure to them if they continue as they have always done with strict attachment to the welfare of his Majesty, from which will depend the following favourable events, viz.

His Majesty has offered to support the rights of the inhabitants to their real property, and until this is ascertained I am bound to keep possession of this country, as likewise until we are sure that the Indians will be pacific.

Contrary to the general expectation, the same indulgence that has until now protected the inhabitants from distress, will be continued during his Majesty's sovereignty in this country, and this being the season in which the planters are employed in the preparing for an ensuing crop, none shall be disturbed from that important object on account of their depending debts.

The misconstruction of what is meant by the enjoyment of the liberty of conscience is hereby removed, by explaining it positively to be, that no individual of this government, shall be molested on account of religious principles, and that they shall not be hindered in their private meeting; but no other public worship shall be allowed, but that generally established in all his Majesty's dominions, which is the Catholic religion.

These important objects that until now have not been published, though resolved, I make known to the public, apprehensive of the dangerous insinuations of several persons, who have made it a business to dazzle the public with false notions, to serve their own purposes in their speculations upon the lands that are lawfully held by all the inhabitants of this government.

I therefore firmly rely, that no person will deviate from the principles of adhesion to our government, until the negociations that are now on foot between his Majesty, and the United States of America, are concluded, and thereby the real property of the inhabitants secured.

Given under my hand, and the seal of my arms, and countersigned by his Majesty's secretary for this government, at government house this 29th day of March 1797.
(Signed) Manuel Gayoso de Lemos.
(Counter-signed) Joh. Vidal, Sec.

Seal.

This proclamation the Governor pretended was issued to quiet the minds of the people, but it had a contrary effect: it was nevertheless well calculated to answer to the purpose really intended; which was that of attaching two powerful classes of the community for the present, (at least) to the Spanish government: these were *first,* the holders of real property, whose cases were not recognized, and provided for by the Treaty, and *secondly,* those in debt. But the people had been so often deceived, under the arbitrary government which had been exercised there, that they had lost all confidence, and the proclamation served more to irritate than conciliate. As soon as this was discovered by the Governor, he requested William Dunbar, Esq. and Mr. Philip Nolan, to inform me that he had just received directions from the Governor General the Baron de Carondelet to have the artillery, and military stores, immediately removed from the forts which were to be given up to the troops of the United States, immediately on their arrival.

The truth of this information was generally doubted, although great pains was taken by the Governor to circulate the report. In order therefore to draw him into a direct explanation, the following letter was sent to him.

Natchez, March 31st, 1797.

Dear Sir,

I was last evening addressed by a number of respectable inhabitants of this district. They are very much alarmed for their situations, in consequence of having expressed their satisfaction since my arrival at this place, of very soon becoming citizens of the United States: but your proclamation of the 29th inst. they conceive renders that event very doubtful. They have therefore from consideration of personal safety, and to avoid the insults which many of them have experienced from one or more officers of a small grade in this district, called upon me to use my influence with your Excellency, to grant to them, and all others who are inclined to leave this country, the privilege of disposing of their property, and passports to enable them to reach the frontiers of such states as they may be inclined to remove to.

I have now stated the substance of their application, and assure your Excellency, that from the respectability of the applicants, it is a subject in which I feel myself interested, and to which I request your Excellency's attention.

Ever since I arrived in this district, I have uniformly recommended to the inhabitants a quiet submission to the government now in force; but at the same time, they have in the most explicit manner, been informed that the period could not be far distant, when the jurisdiction of the United States would be extended to them: but they are not satisfied, they have their suspicions, and it is your Excellency along who can quiet them. Let the cannon, and military stores be again taken out of the fort, withdraw your objections to the descent of the American troops, and their apprehensions will subside. I do not pretend to say that their apprehensions are well founded, it is possible they are not, but your objections to my escort being stationed with me, your hauling back, and remounting the cannon at this place, your dispatching Capt. Minor to delay the arrival of the American troops at this place, added to your proclamation however well meant have had a contrary effect.

I have enclosed two paragraphs from the address, which was handed to me last evening.

I am, Sir, with esteem,
Your friend and humble servant,
Andrew Ellicott.

His Excellency Manuel
Gayoso de Lemos.

To this letter the Governor returned the following answer:

Natchez, March 31st, 1797.

My Dear Sir,

I have just now received your favour of this day, in which you inform me of the application made by several respectable inhabitants of this government to you, requesting your interposition to facilitate to them a privilege that they never ceased to enjoy, and in which consists the greatest liberty of a Spaniard. There is not one single example in our government, of having made opposition to any person selling his property, and leaving the country whenever they called for a passport, and as our system is not altered, I shall not refuse the same privilege to any person that may apply to me for it.

I am very sorry that the persons that addressed you, have imposed on your credulity, and goodness, making use of remonstrances proper to make sensations of the feelings of a good citizen of the United States; but there is not a word of truth in what they have advanced. I have not taken any notice of the satisfaction that some persons have expressed on the prospect of becoming citizens of the United States, nor has any body been apprehended for it, nor have I issued any order for such a purpose, but against Mr. [Thomas] Green, senr., who made his escape conscious of the criminality of his conduct, which is notorious; indeed in all the extent of this government there is but one single individual confined, and that is for a criminal proceeding. There is not a single patrol out in pursuit of any body, nor just in this moment do I find occasion for it; but if I should, I would employ every possible means in my power to suppress disorders, and to keep the peace of the country as I have already done.

I doubt not of the assurances you pretend to give me, of the good advice you have uniformly given to the inhabitants, it being conformable to a gentleman of your character, and whose object is another than that of interfering in the affairs of government.

My proclamation I found absolutely necessary to calm the minds of the people, stating to them, the true situation of the political arrangements between His Majesty, and the United States of America; which does not dissolve the treaty, but requires an essential explanation, not only with regard to the points alluded to in my proclamation, but likewise as I am authorized to declare to you, that the General in chief of the province finds himself under the necessity to consult His Majesty concerning the manner in which the posts are to be evacuated, as it appears from General [Anthony] Wayne's communication to him, that he expects the posts will be delivered with the buildings standing as they are, and by the treaty we conceive that the posts are to be demolished before we quit them, and as such explanations of the true meaning of the treaty either one way, or the other, it might produce unnecessary ministerial contests. My General has given me positive orders, to suspend the evacuation of the posts until the matter shall be amicably settled between the Courts. In the mean while, if the troops of the United States that are daily expected [to] arrive, they shall be received at the Walnut Hills, in the most friendly, and hospitable manner, as is due to a nation with whom we are at perfect peace, and with whom we wish to keep the most perfect harmony.

I flatter myself, that you will do me the justice to acknowledge the propriety of my conduct, in obeying the superior orders of my General, who is actuated by the principles of the strictest honour in supporting the interests of his Majesty entrusted to him.

The uniform good harmony that we have promised reciprocally to one another, will subsist, and it will not only be our duty but likewise our glory, to banish every shadow of misunderstanding which is wrongfully interpreted by the public, without any more foundation than asserted by those that tried to persuade of wrongs that they never suffered.

I am with the most sincere sentiments
of esteem and true friendship,
My dear sir,
Your most humble servant,
Manuel Gayoso de Lemos.

Honourable Andrew
Ellicott.

This letter of the Governor's convinced me of the truth of the confidential communications which I have already mentioned to have received on my arrival, and to bring him to an official, unequivocal explanation for his conduct, was the principal design of the preceding correspondence.

The true cause of the delay being now vowed, the correspondence with the principal part of the foregoing detail, was forwarded with all possible expedition by a young lawyer by the name of Knox, to the Secretary of State, that [the American] government might not be ignorant of our situation, and be enabled to take such prompt measures as the nature of the case appeared to require.

The securing the country to the United States, and the safety of such of the inhabitants as had appeared openly in our favour, were objects of no small importance. Some of the inhabitants had been imprudently warm, among whom was Mr. Green, senr. the same person mentioned in Governor Gayoso's last letter. A very short time after my arrival, he offered me his assistance with that of a hundred volunteers, who he assured me would join in an enterprize, to take the Spanish fort: But he was not the only one who offered his services, they were more general than could be reasonably expected. It was Mr. Green's misfortune that he could not keep his own secrets, and his conduct became known to the Governor, who ordered him to be apprehended; but he made his escape, and retired to the state of Tennessee a few days after I received the Governor's letter of the 31st of March. But of all the propositions that were made to me, that of Anthony Hutchins was the most extraordinary; it was no less than to take the Governor by surprise, and convey him a prisoner into the Chickasaw nation. This proposition was rejected, but not with contempt. Mr. Hutchins was popular with one class of the inhabitants, and it would have been improper to offend him, as his popularity might possibly be made use of to our advantage. This proposition he afterwards made to Lieutenant Pope, but it met with the same fate. The attempt would have been wholly unjustifiable unless in case of open hostility, and being of so singular a nature, it rendered the design very suspicious, and he being at that time a British officer, it did not appear improbable but he had the interest of that nation in view. These propositions were made by Mr. Hutchins a few days after his interview with a confidential friend of Mr. [William] Blunt's! the mysterious gentleman already mentioned.

The alarm was now so great, notwithstanding the professions of the Governor, that it was with difficulty the people could be prevented from acting offensively, and that a general commotion in favour of the United States would take place in the course of a few weeks was evident; the difficulty was, how to direct its effects to the advantage of our country, without committing our government. The attempt was made, and the public is left to judge of its success.

Any further correspondence with the Governor on the subject of my mission now becoming unnecessary, our views were directed to the increasing of our strength. For this purpose the officer commanding my escort, (which was stationed less than two miles from my encampment), inlisted a number of recruits; but none of them could be considered subjects of his Catholic Majesty. On this subject the Governor sent me the following letter.

Natchez, 13th April, 1797.

My Dear Sir,

I am informed that the officer commanding your escort, has enlisted several persons resident of this government, which being against the laws of nations, it cannot pass unnoticed, it being an infringement on the sovereignty of the King, my master, and a disregard of the authority residing in me.

I cannot persuade myself that it was done intentionally, nor thinking that it could give the most remote offence; but as the matter is of the most delicate nature, I request you, to give the necessary orders, that men so enlisted may be discharged, and delivered to Captain Minor whom I commission for this purpose.

The object of the escort not being to raise men in this country while under his Catholic Majesty's dominions; I request of you likewise, to give the most precise, and positive orders to the officer of the troops, or whom it may appertain, to discontinue such proceedings, or any thing that may injure the immunity of the King's dominions, or any of his rights.

The most perfect harmony, and friendship subsisting between his Catholic Majesty, and the United States of America, and the same being recommended in a very particular manner to the individuals of both nations, it would be unaccountable if we, that have the honour to be distinguished by our appointments, did not promote this friendly reciprocity, which not only consists in a hospitable, and polite intercourse, but in guarding, and keeping to one another the prerogatives, and privileges, are due.

Enclosed I have the honour of transmitting to you the list of the men that to my knowledge, have been recruited in this government by the officer commanding your escort.

I have the honour to be,
With the sincerest friendship,
My dear sir,
Your most humble servant,
Manuel Gayoso de Lemos.

Honourable Andrew Ellicott.

To which the following answer was returned.

Natchez, April 13th, 1797.

Dear Sir,

Your Excellency's favour of this date had been handed to me by Captain Minor, but the request it contains, is of so general, and important a nature, and affecting so deeply the privileges of the citizens of the United States, that I must take a short time to investigate its ultimate tendency; as part, if not all the persons named in your Excellency's list, cannot be any construction of the late treaty, or the laws of nations, be considered as subjects of his Catholic Majesty. You may rest assured, that having in view the sacred, and honourable principles, which are the basis of that government which I have the honour to serve, and by which treaties are considered the most sacred of all obligations, I shall be careful neither to infringe the rights of the subjects of his Catholic Majesty, nor willingly suffer an infringement of those of the citizens of the United States.

I have the honour to be,
With great esteem and respect
Your sincere friend,
Andrew Ellicott.

His Excellency Manuel
Gayoso de Lemos.

Here the discussion on this subject ended. The men were neither given up, nor dismissed.

Several of our soldiers at Bacon's landing became unwell about the beginning of April, owing to their disagreeable situation on the bank of the river, being almost surrounded by the annual inundation. They were removed to high ground, about one mile and a half from the river, and the same distance from my own camp.

Having heard nothing from the troops said to be descending the river under the command of Lieutenant Pope but from the Kentucky boats, which occasionally stopped at Natchez on their way to New Orleans, I became very anxious to know their real situation; but my anxiety was released on the 17th of April by a letter from the Lieutenant, by which I found that he had halted at the Walnut Hills in consequence of Governor Gayoso's letter to him.

Immediately upon receiving Lieut. Pope's letter, the following was written, and sent off in the night by a confidential, active person, but the distance being considerable, and the person having no passport, there was more than a probability of his being taken up, searched, and the letter found and opened; it was therefore necessary to be somewhat guarded; but the person was directed to inform Lieutenant Pope, that it would be proper to leave the Walnut Hills, provided it could be done without blood-shed, and to ascertain this fact, he must make the attempt.

> Natchez, April 14th, 1797.
>
> Sir,
>
> Your favour of the 13th has just been received. It is my opinion that the proper place for yourself, and detachment to be stationed, is at this post—here you can be of more service to the United States than at any other place on the river. Nine tenths of the inhabitants, whom I found numerous beyond my expectation, are firmly attached to the United States; but until your arrival, have no rallying point, in case of a rupture between the United States and his Catholic Majesty, which from the conduct of Governor Gayoso I am under the necessity of concluding cannot be very distant.
>
> For particular information I must refer you to the bearer,
>
> And am with respect,
> Your friend and humble servant,
> Andrew Ellicott.
>
> Lieutenant Pope.

After sending off the above letter, doubts began to arise as to the propriety of bringing the troops in the settlement without the approbation of the Governor, at least it appeared best to have it done by his consent: In consequence of which I waited upon Capt. Minor early the next morning, and requested him to be so good as to inform the Governor that I wished to see him at my tent at nine o'clock of that morning. The Governor attended at the time, when Lieutenant Pope's situation was represented to him—that being a military man, with a separate command, ordered by his superior to perform a certain duty, he had no choice left, he must perform, attempt, or resign: that Lieut. Pope had shewn on the present occasion, a spirit of accommodation which was not expected, and perhaps not strictly justifiable: But come, or attempt it, he must, and that being the case, it was certainly better that it should be done in peace, than provoke hostilities by meeting with opposition. The Governor after a little hesitation agreed that the Lieutenant, and his detachment should proceed down the river without opposition.

An express was immediately dispatched at our joint expense to the Commandant at the Walnut Hills, with directions to permit the troops to leave that place without opposition.

I had some expectation that the second express would arrive as soon, if not sooner, then the one sent off the preceding night; which might be the means of preventing disagreeable consequences, if the opposition to the troops leaving that place should extend to force; and to increase the expedition of the second express, I made an addition to pay on my own account: And though he left the town twelve hours after the first, (who was on foot,) the first arrived but about one hour before him—and delivered the letter to the Lieutenant, who immediately began to prepare for his departure. The commandant after protesting against his proceeding, retired into the fort: at this critical moment the second express arrived, and delivered the Governor's orders to the Commandant, who immediately hastened to the landing, and with great satisfaction informed the Lieutenant that he had that moment received orders from the Governor, to permit him to proceed down the river.

On the 24th of April in the forenoon, the detachment arrived at the landing at the town of Natchez, where they remained till the next morning; in the mean time arrangements were made for my escort to leave its encampment at a certain hour the next morning, and join the detachment under the command of Lieut. Pope at the north end of the town. The two companies met in most excellent order, with the colours flying, and attended with their music, and after the usual salutations marched a short distance into the rear of my tent, and encamped on a commanding eminence, having both the fort, and government house in full view.

This junction of the two detachments was neither expected, nor intended by the Governor, who saw with extreme chagrin the whole parade; but it was now too late to provide against it. This measure, with the good appearance of the men, produced such a confidence in our friends, that they entertained but little doubt of our being able to keep possession of the country.

Lieutenant Pope's descending the river was certainly a fortunate circumstance for the United States, though in doing it, he did not strictly comply with his orders from General Wayne, by whom he was instructed to remain at fort Massac, till he obtained some information respecting the evacuation of the posts. And if a judgment was to be formed from the provision made for the detachment, it could not be supposed that it was really intended to descend the river. It was in want of artillery, tents, money, medicines, and a physician. In consequence of this omission, or bad management, I had to furnish the men with such articles as they were in need of, out of the stores appropriated for carrying the treaty into effect: and after all that I was able to do, we had, (to our great mortification) to borrow some tents from the Governor.

From the arrival of the detachment under the command of Lieutenant Pope, till the first of May, there was a perfect silence respecting our business, but on that day the following letter was received.

> Sir,
>
> I have the honour to acquaint you, that the Commander General [Carondelet] of this province, desires me to inform you that his Majesty's Envoy in the United States [Marquis de y Riujo], had given him the intelligence of an attack proposed against our part of the Illinois by the British from Canada, and as such an expedition cannot take place without passing through the territory of the United States, said Envoy did officially communicate what was necessary to the Secretary of State of the United States, requiring that convenient orders should be issued, to have their territory respected, and provide for their own safety, which we doubt not but the United States will acquiesce to, in consequence of the treaty, and good harmony that subsists between the United States, and his Catholic Majesty.

The said Commander General of this Province in consequence of the foregoing information, finds himself under the necessity of putting in a state of defence several points of this river, and particularly [Fort] Nogales (Walnut Hills) to cover Lower Louisiana, in case the British should succeed in their project against Illinois, for which purpose a convenient force shall be sent to Nogales to repair, and defend that post, which far from being against the interest of the United States of America, will in case of being agreed to, leave the military post in that state which it may be found.

As this is a powerful reason, in addition to those that I have offered to suspend the evacuation of the posts, and of running the line, as our attention is entirely drawn towards the defence of the Province.[20] The said Commander General orders me, to pass to you this official communication, and in consequence of the unavoidable delay, to repeat to you in his name, the proposal of remaining here, to go down to Lower Louisiana, or as he thinks might be preferable to remove to Villa Gayoso, where there are sufficient buildings to accommodate you. This insinuation, being an effect of the desire we have to shew every degree of consideration, as a proof of our disposition to improve the friendship between our nations, assuring you that in any part that you should determine to remove to, or stay, the Commander General will facilitate every conveniency in his power for your satisfaction.

I have the honour to be,
With the highest consideration,
Your most humble servant,
Manuel Gayoso de Lemos.

Honourable Andrew
Ellicott.

20. Information on rumored British plans reported by Carlos Martinez, Marquis de y Riujo, Spanish minister in Philadelphia, to Baron de Carondelet, along with circulation of news events regarding Senator Blount's conspiracy, gave credence to both Carondelet's and Gayoso's fears of an impending British assault. Such rumors were rampant in 1797. Ellicott concluded at one point: "A report, believed by all the officers of this [Spanish] Government, that the Republic of France has declared war against the U.S. and that 2,000 Spanish troops have just arrived at Havannah [Cuba] from whence they are to proceed in a few days to this country, to reinforce, and occupy the different posts on the Mississippi" (Ellicott to Secretary of State, Natchez, May 27, 1797, in "Southern Boundary," vol. 1).

A reply of considerable length was immediately drawn up to this letter and communicated to Lieut. Pope, who had likewise received a similar one from the Governor. The Lieutenant objected to my reply, as he thought it would with more propriety come from him, some paragraphs of which he made use of. Upon hearing his objections, which appeared to have some weight, I acquiesced, and only sent the following short reply.

Natchez, May 2nd, 1797.

Dear Sir,

Your Excellency's favour of yesterday is now before me, but as it principally concerns the commanding officer of the troops of the United States in this quarter, who I presume will give a satisfactory answer, it will be unnecessary for me to make any remarks upon it.

In a former communication, you were apprized of my determination to remain at this place until we proceed to the tracing of the boundary, or recalled by the executive of the United States.

I am, Sir, with great respect and esteem,
Your friend,
And humble servant,
Andrew Ellicott.

His Excellency Manuel
Gayoso de Lemos.

On the 2nd of May Lieut. Col. Guillimard, the surveyor appointed on behalf of his Catholic Majesty under the late treaty arrived; I was then informed, that the Governor would be ready to co-operate with me in a few days, but in this report I had no confidence.

On the 3rd of May a number of labourers, and artificers, were engaged in repairing the fort, and several more pieces of artillery were mounted. On the 7th a reinforcement of about 40 men arrived. On the 9th Lieut. Col. Guillimard, and a number of officers, with a boat load of intrenching tools, left the town for the Walnut Hills. Of the truth of Mr. Guillimard's departure I was not certain until the morning of the 11th, when the following letter was immediately sent to the Governor.

Natchez, May 11th, 1797.

Dear Sir,

I am informed, that Mr. Guillimard, the surveyor on behalf of his Catholic Majesty, has left this place for the Walnut Hills. From this circumstance, and the delay on your part, in carrying the late treaty between the United States, and his Catholic Majesty into effect, I feel myself impelled, by the duty which I owe to my country, to request a definitive answer from you, as to the time you will be ready to proceed to the determination of the boundaries between the two nations, agreeably to the stipulations contained in the treaty above mentioned. The harmony, and friendly intercourse between the nations we have the honour to serve, can only be maintained, while they promptly fulfill their contracts with each other.

I have the honour to be,
With great esteem,
Your friend and humble servant,
Andrew Ellicott.

His Excellency Manuel
Gayoso de Lemos.

To which the Governor returned the following answer.

Natchez, 11th May, 1797.

Sir,

I have received your communication of this day, in which you express to me, that Lieut. Col. Guillimard's absence to [Fort] Nogales impelled you, to demand of me, a definitive answer as to the time that I should be ready to proceed to the determination of the boundaries between our nations, in compliance with the late treaty.

Mr. Guillimard's absence is not to be long, and it was in consequence of the suspension that I announced to you the 1st instant, as his presence was essential here for the present.

By my said communication of the 1st instant, you will find that it is out of my power to give you the definitive answer that you require, as it depends on the ministers of both nations to whom this business is intrusted, and through which channel both you, and the commander General of this Province, will be informed of the time that the boundaries are to be determined.

I have the honour to be,
With the highest esteem,
Your most humble servant,
Manuel Gayoso de Lemos.

Honourable Andrew
Ellicott.

On the 16th of May, a company of [Spanish] grenadiers arrived at Natchez, from New Orleans, and after staying about 36 hours, proceeded up the river to the Walnut Hills.

By this time we had received satisfactory information from the Chickasaw, and Chocktaw nations of Indians, that they had for 8 months past been tampered with by people said to be Spanish agents, to oppose the demarcation of the boundary; they were told that immediately upon the establishment of the line, the United States would take possession of all lands on the north side of it, and drive them by force, if they made any opposition. To counteract the effects of these improper suggestions, negotiations were immediately set on foot with the Chocktaws, from whom we had the most to dread, being equally as brave as the Chickasaws, much more numerous, and residing as it were in one neighbourhood. On this subject I shall be more particular in another place.

Mr. Philip Nolan whom I have already mentioned, had now been some weeks in New Orleans; he had at different times been much favoured by the Spanish government, particularly in being permitted to take, and dispose of wild horses, which are to be found in vast numbers west of the Mississippi: and from his singular address, and management had much of the Governor General the Baron de Carondelet's confidence, who informed him, (Mr. Nolan) that the troubles were becoming serious up the river, (meaning Natchez) but that he was determined to quiet them, by giving the Americans lead and the inhabitants hem [hanging]; and asked Mr. Nolan, if he would take an active part in the expedition, to which he replied, 'a very active one.'

The Baron had carried his plan so far, as to direct a camp to be marked out at Baton Rouge for a considerable body of men [2,500], and a contractor was engaged to supply the provision. This intelligence was conveyed to me through a confidential channel, but a knowledge of it was kept from the inhabitants of the district, *first,* because its being known would injure, if not ruin Mr. Nolan, and a few others, and *secondly,* had it been made public, it would have been impossible to restrain some of the inhabitants from committing hostilities. It was thought best, to counteract secretly the plans of the Baron in the city of New Orleans, and turn his weapons upon himself should he persevere in executing his design.

Having heard nothing respecting the intended British expedition from Canada, but from the officers of his Catholic Majesty, it appeared to us, that it was probably no more than a pretext for putting the country in such a state of defence, as to render our getting possession of it difficult, and expensive, and enable his Catholic Majesty to retain a part of it by the negotiation which Governor Gayoso informed us officially was then carrying on.

From every circumstance which came to our knowledge, it appeared certain that there was no design on the part of the officers of his Catholic Majesty to co-operate with us, in carrying the treaty into effect, and the discussion being already carried to a great length, in which we constantly found evasion and finesse, substituted for argument, the correspondence was for the present closed on our part by the following letter to Governor Gayoso.

Natchez, May 16th, 1797.

Dear Sir,

Any observations on your favour of the 11th instant, in reply to mine of the same date, appear to be wholly unnecessary as it contains a specific declaration, that the business which brought me into this country is suspended, and the execution of it, to depend upon future negotiation. I shall now take leave of the subject, by laying before your Excellency a retrospective view of some part of our correspondence, and state some facts which occurred antecedent to your Excellency's letter of the 1st instant.

In your Excellency's letter to me of the 17th of February we find these words, 'I do not conceive, that the least difficulty can arise respecting the execution of that part of the treaty in which you are the actions person.' The day after my arrival at this place, we agreed to proceed down the river on the 19th of March last to meet the Baron de Carondelet, and commence our operations: But previous to that day, I was given to understand by yourself, that the Baron could not attend, and the business devolved upon you. In my communication of the 23rd of March, you will find after mentioning your remounting the cannon at this place, these words, 'and the delay in the business upon which I came, concur in giving me reason to suppose, that the treaty will not be observed with the same good faith, and punctuality by the subjects of his Catholic Majesty, as it will by the citizens of the United States.' In your reply of the same date, we find the following expression. 'The unavoidable detention that has been experienced in beginning the line, you know the reasons; but they shall be removed as Lieut. Guillimard is far on his way up, and at his arrival this important business shall be begun: I do assure you, that there is nothing can prevent a religious compliance of the treaty.' Although your Excellency here declares, 'that Lieut. Guillimard was far on his way up,' he was certainly at home, in New Orleans, sixteen days after the date of your Excellency's letter. In your letter to me of the 25th of the same month you say, 'by every report you are acquainted with the confirmation of every thing I have told you concerning the business, you know that Lieut. Col. Guillimard 'will be here very soon, and that shall proceed immediately to the running of the line.' On the same day you wrote a letter to Lieut. Pope, which was shewn to me by your direction, in which you informed him that the evacuation of the posts was going on, and would be completed in one month. At this very time, you were remounting the artillery in the fort at this place, and rendering it more defensible; of this circumstance you could not suppose me ignorant, as I had written to you upon that subject only two days before. On the 29th of the same month you issued a proclamation, which contained certain evidences that you did not intend strictly to carry the late treaty into effect. This proclamation created considerable discontent, its design was too obvious, to be hidden by the specific pretence, of delaying the execution of the treaty for the purpose of securing to the inhabitants of the district of Natchez their real property. This finesse not succeeding, I was waited upon by William Dunbar, Esq. and Mr. Nolan, who informed me, that you had declared to them, you had just received directions from the Baron de Carondelet, to evacuate the posts within the United States, then occupied by the Spanish troops, and deliver them to the troops of the United States immediately on their arrival; and that boats were then on their way from New Orleans to transport the artillery and military stores, into the territory of his Catholic Majesty. Mr. Dunbar waited upon you the same evening, and gave you to understand that he had communicated the intelligence to me. However respectable the characters who gave me this intelligence, and the source from whence they had it, I entertained strong doubts of its authenticity; and therefore to draw from you its certainty, sent my communication of the 31st of March. In your reply of the same date is the following expression. 'My General has given me positive orders, to suspend the evacuation of the posts, until the matter shall be amicably settled between the two courts.' This confirmed my suspicions. I am wholly unacquainted with any 'matter' to be 'settled,' or any negotiation carrying on 'between the two courts.' And I think I may with confidence say, the Executive of the United States had no idea on the 27th of last March, that his Catholic Majesty would decline carrying into effect a treaty which he had so lately, and with a deliberation of near six months finally ratified: especially when we recollect, that the Court of Madrid, has been remarked beyond any other Court in Europe, for the good faith it has observed, in complying with its national contracts.

From the contents of your last letter, and having no definitive evidence that any Commissioner is authorized to co-operate with me in ascertaining the boundary between the territory of the United States, and that of his Catholic Majesty, any more correspondence between us upon that subject, will for the present be unnecessary.

I have the honour to be.
With great esteem and regard,
Your friend and humble servant,
Andrew Ellicott.

His Excellency Manuel
Gayoso de Lemos.

To this letter the Governor returned the following answer.

Natchez, 17th of May, 1797.

Sir,

I have received your communication of yesterday, by which you are pleased to signify to me, that you find unnecessary the continuance of our correspondence, as you are not persuaded that any commissioner is authorized to co-operate with you, ascertaining the boundaries between the territory of his Majesty and that of the United States. I have informed you, that I am the Commissioner in behalf of his Catholic Majesty, and I have likewise communicated to you the reasons, that have caused the suspension of our business. I still subsist under the said character, but I am not authorized to proceed, until I receive further orders, which I doubt not will reach you, and me, at the same time: therefore as matters are situated, I shall not trouble you with justifying the motives that have caused some disagreement in my communications; they are far from being insincere, but they have originated from the various, momentous occurrences, influenced by the distance from New Orleans, to this place, and time will evince that our conduct is irreproachable.

I have the honour to be,
With the greatest friendship,
And attachment, and respect,
Your devoted, humble servant,
Manuel Gayoso de Lemos.

Honourable Andrew
Ellicott.

On the 19th of May, more troops passed Natchez on their way to the Walnut Hills. The reinforcing of that post, and the fort at Natchez, kept the inhabitants in constant fear; they considered the whole as designed against them, and in order to avert, what they conceived would shortly become a serious calamity, a number of plans were now devised; but they were objected to as being premature, and having a tendency to involve the United States in a war. The inhabitants were constantly told, that self defence ought to be their object, and to make that effective, they ought at all times to be ready to repel an attack, and if ready, there could be not a doubt of their success.

One of those plans for attacking the Spaniards was of a very general nature, and covered many sheets of paper; it was drawn up by a Mr. [Jonathon] Dayton, a native of the state of Connecticut, but being a suspicious character, his plan was detained and put into the hands of George Cochran, Esq. to be used against himself, if he should be found playing a double game, or joining the Spanish interest.

About this time a serious difference took place between the Governors, the Baron de Carondelet, and Gayoso, which no doubt must have embarrassed their proceedings, and weakened, if not entirely disconcerted their plans. A number of conjectures were made to account for this difference; but none of them hit upon the true cause. Upon this subject I received the following letter from Governor Gayoso.

Concord, May 28th, 1797.

Dear Sir,

I have the honour to address you on a subject, that though it does not concern our public business, it is connected with my conduct towards you. I find the Commander General of this Province, is informed that there has passed several occurrences between you and me, of which I have not given him any notice. I cannot recollect any of our transactions that is not either expressed, or alluded to in our correspondence. With regard to my intercourse with you merely as a gentleman, I hoped it only belonged to us to judge of, but yet it might be such, through inadvertency, as may have drawn the attention of the public, who sometimes misplace the man from his private, to his public character; but even in this case it is not in my principles to act any otherwise than as a gentleman.

I request you therefore, to remind me of any particular circumstance that is not comprehended in the heads that I have expressed, and to be very particular to remark if my conduct towards you, either in a public or private character, has given you the least cause of displeasure.

I have the honour to be,
With the highest esteem and regard,
Your most humble and obedient servant,
Manuel Gayoso de Lemos.

Honourable Andrew Ellicott.

This letter of Governor Gayoso's appeared in some measure calculated to make me a party in the dispute, the appearance of which I was determined to avoid, and therefore endeavoured to be as guarded as possible, and made no reply till the 31st, when the following was sent to the Governor.

Natchez, May 31st, 1797.

Dear Sir,

I have the honour to acknowledge the receipt of your Excellency's favour of the 28th. I cannot recollect any transactions between us, which can be considered as official, 'that not either expressed, or alluded to in our correspondence,' in which I presume you have acted agreeably to your instructions, and therefore, (though I lament the delay on your part, in carrying the late treaty between the United States and his Catholic Majesty into effect, as it will have a tendency to disturb the harmony, now subsisting between the nations we have the honour to serve,) I do not impeach your motives.

'With regard to' your 'intercourse with' me, 'merely as gentleman,' it has constantly been such, as became a person in your station, and consequently entirely satisfactory.

I have the honour to be, with great esteem,
And regard, your humble servant,
Andrew Ellicott.

His Excellency Manuel
Gayoso de Lemos.

On the first day of June, I received the following proclamation of the Baron de Corondelet's, by a private conveyance from New Orleans, and not having heard of it from Governor Gayoso, I had some doubts of its authenticity, and called upon Captain Minor to know the truth of it; but it appeared that neither he, nor Governor Gayoso were acquainted with it at that time: it was announced officially two days after, and had an effect different from what the Baron expected.

Province of Louisiana.
By his Excellency Francis Lewis Hector, Baron

De Carondelet, Knight of the order of Malta, Major General of his Armies, / Commandant General of Louisiana, West Florida, Inspector of the Troops, Militia, Etc., Etc., Etc.

Whereas it has come to our knowledge, that some evil disposed persons, who have nothing to lose, have been endeavouring to draw the inhabitants of Natchez into improper measures, whose disagreeable consequences, would only fall on those possessed of property, whilst the perturbaters would screen themselves by flight. We have thought convenient, in order to dissipate the reports, which without the least foundation, are intentionally propagated to alarm the inhabitants, to declare authentically, as we do by these presents, that the suspension of the demarcation of the limits, and the evacuation of the forts which will be comprehended on the other side of the line, is at present occasioned by the imperious necessity of securing Lower Louisiana, from the hostilities of the English, who without regard to the inviolability of the territory of the United States, have set on foot an expedition against Upper Louisiana, which they cannot however attack, without traversing the aforesaid territory, an act as violent, as unjust toward the United States, leaving no room to doubt, that if they make themselves masters of the Illinois country, they would avail themselves of the articles of the last treaty, concluded between the United States and Great Britain, in appearance contradictory to that concluded between the former and Spain, by which they guarantee the navigation of all rivers, on which the said United States hold posts, to attack or molest Lower Louisiana. We have thought proper to put the posts of Walnut Hills in a respectable, but provisional state of defence, until the United States are informed of these motives, by the Minister Plenipotentiary of his Majesty, to whom we have communicated them, to provide against these inconveniencies, and by taking the proper steps to cause the territory to be respected, shall put it in our power to fulfill without danger the articles of the treaty concerning limits.

In consequence we indulge the well grounded hope, that the inhabitants of Natchez will behave with the same tranquility, and entertain the same affection, of which they have (given) the proofs on occasions towards the Spanish government, which could not, without feeling in the most sensible manner, see itself forced to compel the unsubordinate minds, to hear the dictates of national gratitude in a settlement, which it has been at so much pains and expense, to form and protect.

And, that the present proclamation may get to all, and every one in particular, we order that it may be published, and posted up in all the districts of the government of Natchez in the usual places.

Given under my hand and seal, at the house of government,
In New Orleans, the 24th of May, 1797.
(Signed) Baron De Carondelet.
(Counter-signed) Andres Lopez Armestro.

It has already been remarked, that the foregoing proclamation was not attended with any beneficial effects to the Spanish government: on the contrary, it served to convince the inhabitants, that his Catholic Majesty intended if possible, to retain the country under one pretence or other till the treaty should become a dead letter. The most common observer, could not fail to contrast the reasons given by the two Governors, in their proclamations for retaining the posts. Governor Gayoso by his proclamation of the 29th of March, informed the inhabitants, that it was for the purpose of having their real property secured to them: and the Baron de Carondelet, that it was for the purpose of opposing the British, who were meditating an attack upon that country. The Baron was much more unfortunate in the motive he assigned for keeping the posts, than Governor Gayoso, and much less acquainted with the political sentiments and prejudices of a large class of the inhabitants, who had formerly been British subjects, and to which government many of them were still attached; both from principle and habit, and no intelligence could have been so pleasing to them, as that of the British preparing to repossess that country: in this business the Baron was guilty of committing a very glaring oversight for a person of his extreme caution, and sound judgment. He appears extremely anxious, that the United States should cause their neutrality to be respected, if the British should attempt to pass through any part of their territory, to attack the possessions of his Catholic Majesty; but at the same time, insists upon holding posts within the United States, to oppose the same power upon the principle, the British would have an equal right to complain of our submitting to it.

After the appearance of the Baron's proclamation, the public mind might be compared to inflammable gaz; it wanted but a spark to produce an explosion! A country in this situation, presents to the reflecting and inquisitive mind, one of the most interesting and awful spectacles, which concerns the human race.

About the beginning of June, an itinerant Baptist minister by the name of [Barton Hannon] Hannah, called upon me, and asked permission to preach a sermon in my camp the following Sabbath, which was the 4th of the month. As no public worship was allowed in any of the Spanish colonies, but that of the Roman Catholics, there appeared to be a difficulty in complying with his request; but to get clear of this obstacle, and to give no just cause of offence to the officers of his Catholic Majesty, I spoke to the Governor upon the subject; who gave his consent without hesitation. On the morning of the 4th, it was stipulated with the preacher, that he should not touch upon political subjects in his discourse, that the public mind was already extremely agitated, and that a few observations however just, and well intentioned, might be productive of great mischief.

A public protestant sermon, being a new thing in that country, drew together a very large, and tolerably respectable audience: and though the preacher meddled not with politics, the effect was nearly the same; the hearers who were generally protestants, wanted liberty of conscience in its fullest extent, and very naturally preferred a sermon which they understood, to a mass, which few of them knew any thing about. The preacher being a weak man, was extremely puffed up with the attention he received on that occasion, which arose more from the novelty of the case, than his own merit and talents, and paved the way for commotion which took place a few days after.

In case of a rupture with the United States, the officers of his Catholic Majesty calculated largely upon the effects of their intrigues, and the money they had expended in the states of Kentucky, Tennessee, and other districts, west of the Allegany mountains, for the purpose of detaching the inhabitants from the Union.[21] These intrigues so far as I was able to discover, appeared

21. In a confidential dispatch to the Spanish Supreme Council of State (September 25, 1787), Governor Estevan Rodrigues Miro of Louisiana referred to James Wilkinson as "Spy #13," who had been in contact with Miro himself on plans beneficial to Spain. According to Governor Miro, Wilkinson's first proposal was to promote settlers in Kentucky to immigrate; second, he proposed that Spain undertake direct intervention to support the separation of western Kentucky from the Union in order for it to become a new Spanish province. See Gilbert C.

to have originated with Mr. [Diego de Gardoqui] Gardauqui, at the time he was his Catholic Majesty's minister [chargé d'affaires] to the United States. But the advantages to be derived from these intrigues, were certainly very much overrated.

From a conviction founded on the undeniable evidence contained in several authentic documents, of the reality of those intrigues, suspicion naturally attached itself to every equivocal character who was observed to be passing and repassing between those states, and New Orleans. These suspicions in one instance fell upon a gentleman[Thomas Power, known confidant of James Wilkinson], who had been very active in these intrigues, and who I was informed on the 5th of June, was immediately setting out by order of the Baron de Carondelet, for the states above mentioned. The critical situation of the country at that time, warranted a conclusion, that his business was to serve the Spanish government, at the expense of the United States. From this impression, measures were immediately taken to embarrass him in his progress, by writing to certain characters in those states, of whose fidelity to their country, there could be no doubt, to have him detained and examined, as often as possible; the success was a great as could be expected. The secretary of state of the United States [Thomas Pinckney] was likewise written to on the same subject, on the day before mentioned.

My first written communication to the Indians was likewise sent on the 5th of June: till that time, our correspondence with them had been merely verbal. It required some time to discover the inclinations and wishes of the traders and interpreters, and after being fully satisfied on this head, a circular letter was addressed to such of them as appeared to merit confidence, to be communicated to the chiefs. As this subject would probably be very uninteresting at this time, but little will be said upon it; it was however attended with considerable difficulty, and if circumstantially detailed, would of itself require a volume. It was a subject that required great caution in the management, so

---

Din, "Spanish Immigration Policy in Louisiana and the American Penetration, 1792–1803," *Southwestern Historical Quarterly* 76 (1973): 255–76; and, A. P. Nasatir and Ernest P. Liljegren, eds., "Materials Relating to the History of the Mississippi Valley from the Minutes of the [Spanish] Supreme Council of State, 1787–1797," *Louisiana Historical Quarterly* 21, no. 1 (January 1938): 5–75; and Cox, *The West Florida Controversy,* 46–49. Ellicott sent a report on the topic to the Department of State on November 14, 1798. In that letter, General Wilkinson and several others are mentioned as having been in the pay and interest of Spain" (qtd. in Wilkinson, *Memoirs of My Own Times,* vol. 2, appendix 28).

as to give no offence to the officers of his Catholic Majesty. Our whole view, was to render the Indians harmless, in case of a rupture between the United States and Spain, and to convince them that it was their true interest, to take no part in the quarrels between white nations. War of itself, with all the alleviating circumstances introduced by the present refined state of civilization, should be considered as one of the greatest scourges of the human race; why then add to its calamities by employing savages in it?

The success of those negotiations with the Indians was so complete, that in less than three months they were almost wholly detached from the Spanish interest, and although the United States had no treaty with the Chocktaws, through a large extent of whose country we had to pass, they gave us no molestation, in the execution of our business. A copy of every paper and message relative to this subject, was with my other communications forwarded to the secretary of state, where I presume they remain filed.

The afternoon of Friday the 9th of June, I spent at the Spanish commissaries, and after taking tea in the evening, rode to my camp, which I found to be in great confusion, and upon inquiring the cause, was informed that the Baptist minister was taken and sent to prison; but on what account I could obtain no information. From the state of the country at that time, the worst of consequences were to be expected from this measure. After staying about ten minutes in the camp, and recommending in the strongest terms the necessity of keeping good order, I rode to Captain Minor's, and desired him to wait upon the Governor as quick as possible, and request in my name, the liberation of the minister; not as a matter of right, but for the purpose of preserving peace in the country. The Captain lost no time in repairing to the Governor at Concord [government house]; where he found him in state of considerable agitation, and deaf to the request. I remained at the Captain's till his return, and after receiving the Governor's answer, inquired the cause of the minister's confinement, and received the following statement, which upon further inquiry, I found to be correct. The minister being elated with the attention he had received on account of his sermon, and imboldened by having the permission to speak publicly, he had with enthusiastic zeal, which was a little heightened by liquor, entered into a religious controversy in a disorderly part of the town, generally inhabited at that time by Irish Roman Catholics, who took offence at the manner in which he treated the tenets of their church, and in revenge gave him a beating. He immediately called upon the Governor, and in a peremptory manner demanded justice; threatening at the same time to do it for himself, if his request was not complied with. The Governor with more patience and temper than ordinary, desired him to reflect a few minutes, and then repeat his request, which he did in the same words, accompanied with the same threat. Upon which the Governor immediately ordered him to be committed to the prison, which was within the fort, and his legs to be placed in the stocks.

Between ten and eleven o'clock in the evening, I returned to my camp, and went to bed; but spent great part of the night in devising plans, to direct the commotion which was now inevitable, to the advantage of the United States, without committing either the government or its officers.

On the morning of the tenth, agreeably to my constant custom, I rose about daylight, and walked into the town; when to my surprise I found many of the inhabitants up, and the place in considerable confusion; upon inquiring the cause, I was informed that the Governor, with the officers of the government and several Spanish families, had taken refuge in the fort, to avoid the fury of the country inhabitants! Thus in less than ten hours, by an unnecessary, (at that time an impolitic) exertion of power, the authority of the Governor was confined to the small compass of the fort.

This proceeding of the Governor, was considered by the inhabitants an attack upon the privileges of the citizens of the United States, (Mr. Hannah being one,) and a determination at all events to enforce the laws of Spain, both civil and religious with rigour.

The tumult appeared to have no direct object in view, it was without system, or a rallying point. Some were for immediately attacking the fort, and others for taking the gallies, and making themselves masters of the river; but much of the greater part were taking the fort. By the evening of the tenth, the opposition to the Spanish government had become very general over great part of the district.

On Sunday the 11th, a number of the enterprising opposers of the Spanish government, called upon Lieutenant Pope, and myself, and declared their determination of commencing hostilities in consequence of the imprisonment of Mr. Hannah, and the following address of the Baron de Carondelet's, which they considered as a declaration of war against the United States. The address was in those words, viz.

> The government being informed by his Majesty's ambassador to the United States of America, that an expedition assembled on the lakes,

was intended to attack the Illinois, had judged necessary for the surety and tranquility of Lower Louisiana, to suspend the evacuation of the post at Natchez and the Walnut Hills, being the only posts that cover it: the possession of which will put the English in a situation to disturb and ravage the country, in case they rendered themselves masters of Upper Louisiana with so much more facility, as by an article of the treaty concluded posteriorly with Great Britain, the United States acknowledge that the English may freely navigate and frequent the posts belonging to the United States, situated on the rivers in general, lakes, Etc., being a manifest contradiction to the treaty concluded with Spain, which appears to annul, because of this the United States acknowledge that no other nation can navigate upon the Mississippi without the consent of Spain.

Notwithstanding the legitimacy of these motives, the suspension had been represented to the congress of the United States, with all the necessary veracity, and intimated by our orders to the commissary of limits, as well as to the Commandant of the detachment of American troops now Natchez. We are now informed, that a detachment of the army of the United States cantoned on the Ohio, are on their way from Holstein towards Natchez, while the militia of Cumberland are intimated too hold themselves ready, to march at the first notice.

These hostile dispositions can naturally only concern these provinces, because the United States are in peace with all the savages. The anterior menaces of the commissary of limits, and the Commandant of the detachment of Americans now at Natchez, the immediate rupture, (and if the American gazettes are to be believed,) already between France, our intimate ally, and the United States, engage us to be on our guard, to defend jour property with that valor and energy, which the inhabitants of these provinces have manifested on all occasions, with the advantage and superiority which knowledge of our local situation will procure, and with the confidence which right and justice inspire. If the congress of the United States have no hostile intention against these provinces, they will either leave the post of Natchez, or the Walnut Hills, the only bulwarks of Lower Louisiana , to stop the course of the British, or give us security against the article of the treaty with Great Britain, which exposes Lower Louisiana to be pillaged and destroyed down to the capital, we will then deliver up the said posts, and lay down our arms, which they have forced us to take up, by arming their militia in time of peace, and sending a considerable body of troops by round about ways to surprise us.

New Orleans, 31st May, 1797.
(Signed) Baron de Carondelet.[22]

I shall make a few remarks on this address of the Baron's, and then pass on to the subject of the commotion.

The Baron in this address, pursues the same inconsistency as he did in his proclamation; he appears not to be aware of the impropriety of fortifying within the territory of a neutral power, (without previous consent,) to oppose the enemies of his sovereign. He likewise seems to be wholly ignorant of the geography of the United States, or he could never suppose, that a body of troops were preparing to march from the Ohio by the way of Holstein to attack them, which would be an undertaking little short of Ferdinand de Soto's, when he traversed the Floridas. But why march many hundred miles through the wilderness, and consume the best part of a year to attack them, when the same effect might be produced in a few days, by descending the river?

22. Cox, *The West Florida Controversy,* 49–50; Holmes, *Gayoso,* 189–94; and the text "Proclamation of Baron de Carondolet," May 31, 1797, which was sent to the secretary of state, from Natchez (June 27, 1797), along with several other documents (in "Southern Boundary," vol. 1). The documents in this dispatch contain Ellicott's account of the tenseness of the situation in Natchez during June, and the inhabitants' support for the United States: "The business now put on a very serious aspect, and hostilities appeared inevitable; however, I stood alone in the opinion that an amicable & honorable compromise would take place.—By this time the opposition to the Spanish Govt. had assumed some form; a number of respectable militia companies had elected their officers & were ready to take the field. . . . Companies of militia were forming & organizing in the country, and the Governor exerting himself by strengthening & reinforcing the fort. He called to his aid every person who would join him, either through attachment or fear he was nevertheless too weak to attempt any offensive operations" (Ellicott to Secretary of State, Natchez, June 27, 1797, in "Southern Boundary," vol. 1). The "commotion," as Ellicott called it, ended peacefully. In a report later that year to Secretary Pickering, Ellicott reflected on his thoughts and conclusions: "I thought the Spanish government very weak in the district of Natchez when I first arrived—In that I was not mistaken.—I think this equally, if not more so, I am convinced the present government might be abolished by the materials within itself, and that but little risk to those who might undertake it. And what contributes considerably to this weakness, is the general opinion of the inhabitants that it will unquestionably before many years be annexed to the U.S." (Ellicott to Secretary of State, New Orleans, January 13, 1799, in "Southern Boundary," vol. 3).

The conduct of the United States, in admitting the British to participate in the navigation of the Mississippi, I never attempted to justify.

The assertion of the Baron, that I had menaced their government, was without foundation, but it is not improbable that he had received such information: the close of the last paragraph admits that they had 'taken up arms,' which many of the inhabitants of the district of Natchez were of opinion amounted to a virtual commencement of hostilities, if not to a declaration of war: but it is certain this was rather a forced, than a just and liberal construction. But to return to the subject of the commotion.

To have opposed the inhabitants directly in their opposition to the Spanish government, would have put an end both to our influence and power of being useful; and to encourage them, would in my opinion have been improper, as the United States had not extended their jurisdiction to that territory. I therefore on my part, resolved to do neither, but to direct if possible the attention of the people from immediate acts of hostility, by address and management, to other objects, till their proceedings were reduced to system, under which they might be more easily checked if necessary, or their force rendered more efficient, if the officers of his Catholic Majesty should commence hostilities against them.

In pursuance of this plan, the spirit of the people was highly complimented on account of their present exertions, but in the conclusion they were informed, that before they could call with propriety upon the United States for aid or support, in case they were unfortunate, or likely to be overpowered, it would be necessary that the United States should have some evidence of their having made an election in their favour. It was therefore proposed, that they should subscribe a specific declaration for this purpose as generally possible, before hostilities were attempted; but in the mean time, not to lose sight of their personal safety, and to be ready at a moment's warning to act in concert and repel an attack if made upon them. This measure was generally approved, and the subscription paper put into immediate circulation.

On the evening of the 12th, Lieutenant Pope and myself, received a verbal message from the Governor, by his aid Captain Minor, to the following purport, 'gentlemen, Governor Gayoso requests the favour of an interview with you to-morrow, as private gentlemen; an interview to be without the fort, to see if some plan cannot be devised to quiet the present disturbance in the country.' To this message, I replied that I had no objection to the proposed interview, that I approved of peace, and would join in any measures for that purpose, consistent with the honour and safety of the inhabitants, who generally considered themselves citizens of the United States. Lieutenant Pope's answer was the reverse, he refused to have anything to do with the Governor, or his messages, unless delivered in writing. As the message was jointly to Lieutenant Pope and myself, and as the Lieutenant would not agree to the interview, I informed Captain Minor that I should not attend alone, upon which he returned to the fort.

On the 13th I received the following letter from Governor Gayoso.

> Sir,
>
> By repeated informations, and by every appearance, it seems past a doubt that a number of the inhabitants of this government, subjects of his Majesty, are at present in a state of rebellion, with the hostile design of attacking this fort.
>
> I am informed that yesterday, several of the said insurgents were riding through the country, soliciting subscribers to a list that already contained the names of several persons, who declared themselves citizens of the United States of America, though they are actually under oath of allegiance to his Majesty, and under whose dominion and protection they have lived, and enjoyed the benefit thereof, and the bearers of this list declare themselves commissioned by you for that purpose.
>
> I cannot prevail upon myself, to believe that you have either authorized or encouraged such proceedings, as a conduct of that nature would unavoidably produce the most disagreeable and fatal misunderstanding between our nations, and the total destruction of this district.
>
> Therefore I request you, to give me such a positive answer as will enable me, to inform the Commander General of this province for the intelligence of his Majesty, of the part you take in these transactions, and should you take such an active part in them as it is represented you do, from this moment I protest in the name of the said Commander General, against such conduct, and make you answerable for the fatal consequences that may ensue. I repeat the request of a positive answer on this subject.
>
> I have the honour to be, with the greatest regard,
> Sir, your most humble servant,
> (Signed) Manuel Gayoso de Lemos,

Brigadier of the Royal Armies, Governor of Natchez,
and its dependencies, Etc., Etc., Etc,.

Honorable Andrew
Ellicott.
Fort at Natchez, June 13th, 1797.

To this letter the following answer was immediately returned:

Natchez, June 13th, 1797.

Dear Sir,

In order to answer your letter of this day, which from the spirit of it, denies the existence of that principle which has been the object of a long train of discussion between us, I must refer to your letter, date the 12th of March last. In that letter, you admit not only that at Daniel Clark's will be about the point of demarcation, but that the Commissioner of his Catholic Majesty, would in all probability meet me at that place. As the treaty itself was a fact notorious, so likewise ought to be all the transactions attending it, either in direct performance, or open violation. The people therefore became acquainted with those circumstances, that were the result of my observations, or of the acquiescence of the Spanish government. They were matters which involved their felicity, and could not from decency or duty be withheld.

If on the present occasion, the people have thought proper to act in conformity to the intelligence received, which intelligence had the combined sanction of the agents of both governments for its support; is my agency or my conduct, to be called to account with regard to the effects?

A little inquiry into the human mind, would have enabled you, sir, to have discovered a more powerful cause, than any operation of mind concerned in producing the present commotion. The people conceived themselves citizens of the United States, they had a right to conceive themselves so, and they have lately individually come forward to express their wishes and intentions.

After this short detail, of what is the real cause of the present disturbance, I might flatter myself with a complete acquittal on your part, did not the first paragraph of your last letter compel me to form a different conclusion.

On what principle do you still retain the idea, that the citizens of this country are subjects of his Catholic Majesty? Is there not a compact deliberately entered into by the two nations, to the contrary of your opinion? Have not you acknowledged me to be the agent of the United States to carry that compact into effect? And have you not repeatedly pledged your word to co-operate with me in that object? Here I might with propriety ask, what human assurances could have gone further than those that have been made on your part? Do all solemn obligations between nations depend upon chance or caprice, or is there such a principle universally acknowledged among different powers as the law of nations? If your Excellency admits that there is such a principle as national laws, I assert that the inhabitants of this district, cannot be considered in any wise subject to the Spanish monarchy. If you deny the existence of the principle, I have only to observe, that the people cannot with propriety be censured for recurring to that conduct, which will ultimately secure their felicity.

I have proceeded thus far by way of argument, in answer to your communication, from the whole of which you will readily infer a very natural conclusion, that the delay on your part, in carrying the treaty into effect, added to the invariable nature of the human heart, have produced the evils of which you complain, But since you demand a positive reply to the general question, whether I am concerned in measures destructive of his Catholic Majesty's interest, or in an attempt to attack the fort, I give you my honor, that I am not.

As you have assisted me in confirming the sentiment that this territory belongs to the United States, I do now on the part of the United States, (as their commissioner,) solemnly protest against the officers of his Catholic Majesty, landing any troops, or repairing any fortifications in the territory before mentioned, as I shall consider such conduct a violation of the treaty, and an immediate attack upon the interest, honour, and dignity of my country.

I shall now finally observe, that from your verbal message by your aid Capt. Minor, I expected that your Excellency would have proposed some plan of accommodation consistent with the justice and sentiment of the countries we have the honour to serve: and should you have any such proposals to make, I assure you that I feel every wish to enter into a discussion for that purpose.

I have the honour to be,
With great esteem and regard,
Your friend and humble servant,
(Signed) Andrew Ellicott.

His Excellency Manuel
Gayoso de Lemos.

The foregoing reply to the Governor's letter, was sent to him by a flag about four o'clock in the afternoon. About eleven o'clock the ensuing evening, I received a verbal message by George Cochran, Esq. [Natchez planter] from the governor, requesting a private interview with me at the house of the said Cochran, at nine o'clock the next morning. To this I had no objection, as I supposed the object of the interview was to fall upon some plan of accommodation. The next day being the 14th, I met the Governor agreeably to his request. He at first appeared much agitated, and intemperate, and endeavoured to intimidate by threatening to bring the Indians on the settlement, upon which I put on my hat, as I was going to leave him; but he insisted upon my staying a few minutes longer. He was then given to understand, that the interview was in consequence of his own request, that if he had any proposal to make for quieting the disturbances in the country, it would be attended to, and calmly discussed, provided it was neither accompanied by threats, or attempts at intimidation. To which he assented; we then entered on the subject, and mutually agreed upon certain principles, which appeared well calculated to produce the desired effect: but the difficulty was how to bring Lieut. Pope into the measure, without which the whole would be abortive; this was however fortunately effected by the address of Mr. Cochran, and a few other well disposed gentlemen. On the 15th, the Governor published a proclamation, embracing in part the terms we had agreed upon, and containing some expressions very offensive to the people. The proclamation was communicated to me for my approbation previous to its publication; but as it did not completely embrace the objects we had previously settled, my assent was withheld, though an assurance given, that it should meet with no opposition from me. Its fate was such as might reasonably be expected from the state of the country. In some places it was torn to pieces, and in all, treated with contempt. As soon as I discovered that the proclamation, instead of quieting, only served to irritate, and render the inhabitants more violent, I gave the information to the Governor's aid Capt. Minor, but it was not too late to remedy it.

The commotion now assumed a very serious aspect, and hostilities appeared inevitable; but even at this period, it was my opinion that an honourable and amicable compromise would take place.

By this time the opposition to the Spanish government had acquired some form, and more strength: a number of militia companies had elected their officers, and were ready to take the field.

On Friday the 16th it was agreed, that on Tuesday the 20th following, a meeting of the principal inhabitants of the district, should be held at Mr. Belts, about eight miles from the town. The design of calling this meeting, was for the purpose of taking the business out of the hands of the people at large, and commit it to a committee of their own electing. In the mean time both sides continued their preparations, companies of militia were forming and organizing in the country; and the Governor exerting himself in reinforcing and strengthening the fort: he called to his aid, every person who would join him either through fear or affection, but was nevertheless too weak to attempt offensive operations.

On Sunday the 17th, about ten o'clock in the evening, a Spanish patrol fell in with a patrol from my camp, and without hailing or other previous ceremony, fired upon it; the fire was immediately returned, but I believe little or no damage was done. Immediately on the commencement of the firing, all the lights in my camp were extinguished, they might have served a good direction for the Spanish artillery, one piece having been brought to bear upon my tent for a number of weeks. Two or three days after I had first observed it, the impropriety of such conduct was mentioned to Governor Gayoso, who ordered the direction of the piece to be changed, but it was brought back again in a few days.

On the evening of Sunday the 18th, I received a verbal message from the Governor, by his aid Capt. Minor, requesting a private interview with me the next morning, at the house of his aid, to which I immediately consented.

The next morning the Governor left the fort, and by a circuitous route, through thickets and can brakes, made his way to the north side of his aid's plantation, and thence through a corn field to the back of the house, and entered the parlour undiscovered, where I joined him. Our conversation immediately turned upon the state of the country. He assured me that he was sincerely desirous of coming upon some terms of accommodation with the inhabitants, and he understood I was going to attend the meeting at Mr. Belts, requested that I would be so good, as to use my influence in bringing about a compromise. He was told that was my object, and that a plan was already arranged, which would check, and finally put an end to the present disturbance: but that no terms could now be expected, that were not safe and honourable to the people, they had felt and knew their strength, and would only agree to disband and return home, by being admitted to enjoy a qualified state of neutrality till the treaty between the United States and his Catholic Majesty should be carried into effect: to this privilege I thought them entitled, but to go further, would, in my opinion be impolitic, and probably attended with ruin to individuals, if not to the district. As the Governor neither commented on, nor appeared dissatisfied with this observation, it was taken for granted that he would agree to the qualified neutrality proposed. Before we parted, the Governor was informed of my determination, to abide by that declaration contained in my last letter, to protest against the landing any more troops on the east side of the river, above the 31st degree of north latitude, except for the purpose of refreshment; which he ever afterwards observed in the most honourable and punctilious manner.

My feelings were scarcely ever more affected, than in this interview with Governor Gayoso. The humiliating state to which he was reduced, by a people whose affections he had courted, and whose gratitude he expected, had made a strong and visible impression upon his mind and countenance. His having been born and educated with high ideas of command and prerogative, served but to render his present situation the more poignant and distressing.

A circumstance took place on the morning of the 20th, which had a considerable effect in attaching the Chocktaws to the interest of the United States. A short time after our arrival at Natchez, a body of those Indians crossed the Mississippi, to make war upon the Cadoes [Caddo]. In this expedition they were very successful, and returned with a number of poles filled with scalps; but in place of finding the Governor at his usual residence, they had to pay their respects to him in the fort. From this time their respect for him and his people as warriors, was diminished, and their attention to the Americans increased.

After spending a short time with the Indians above mentioned, I went on with a number of gentlemen to the meeting at Mr. Belt's, which was large and respectable. In consequence of the previous arrangements, that had been agreed upon by the gentlemen of property and influence in the country, but little difficulty was found in prevailing upon the people, to submit the future management of their affairs to a committee, to be chosen by themselves. In this stage of the business, Mr. Hutchins took an active, useful and decided part, which to me appeared not only extraordinary, but inexplicable, when I reflected on his proposition to take the Governor by surprise, and his wishes frequently manifest, to remove the Spaniards by force.

About three o'clock in the afternoon, the people proceeded to the election of the committee, when the following gentlemen were almost unanimously chosen.

Anthony Hutchins,	Cato West,
Bernard Lintot,	Joseph Bernard,
Isaac Gaillard,	Gabriel Benoist.
William Ratliff,	

To the above committee, Lieut. Pope and myself were added, by unanimous vote of the meeting.

The same evening the committee assembled in the town of Natchez, and informed the Governor by a note of their election, to which he immediately returned a polite answer.

On the 21st, the committee proceeded to business. The Governor offered them the use of government house, which was then unoccupied, but they declined accepting of it, and made use of one which I was preparing to go into. It was a good new building, nearly finished, the property of Mr. Dunbar, who offered me the use of it, free of expense to the public: upon these terms only did I consent at any time during my absence, to reside in a house.

On the 22nd the following letter and propositions, were presented to the Governor for his concurrence.

Natchez, June 22d, 1797.

Sir,

The following propositions being unanimously agreed to by us, the underwritten, being a committee appointed by a very respectable, and numerous meeting of the inhabitants of this district, and Andrew Ellicott commissioner and citizen of the United States, and P. S. Pope, commanding the United States troops on the Mississippi, are submitted to your Excellency, with a request that you may accede to, and transmit a copy of the same to the Baron de Carondelet, and obtain his concurrence, in order to restore tranquility to this government.

1st. The inhabitants of the district of Natchez, who under the belief and persuasion that they were citizens of the United States, agreeably to the late treaty, have assembled and embodied themselves, are not to be prosecuted or injured for their conduct on that account; but to stand exonerated and acquitted.

2dly. The inhabitants of the government aforesaid, above the 31st degree of north latitude, are not to be embodied as militia, or called upon to aid in any military operation, except in case of an Indian invasion, or for the suppression of riots during the present state of uncertainty, owing to the late treaty between his Catholic Majesty and the United States not being fully carried into effect.

3dly. The laws of Spain in the above district shall be continued, and on all occasions be executed with mildness and moderation; nor shall any inhabitant be transported as a prisoner out of this government on any pretext whatever: and notwithstanding the operation of the law aforesaid is hereby admitted, yet the inhabitants shall be considered to be in an actual state of neutrality, during the continuance of their uncertainty, as mentioned in the second proposition.

4thly. We the committee aforesaid, do engage to recommend it to our constituents, and to the utmost of our power endeavor to preserve the peace, and promote the due execution of justice.

We are your Excellency's most obedient humble
Servants,
(Signed)
Anthony Hutchins,
Bernard Lintot,
Isaac Gaillard,
William Ratliff,
Cato West,
Joseph Bernard,
Gabriel Benoist.

Don Manuel de Lemos, Brigadier
In the Royal Armies, Governor

Military, and Political of the Natchez,
And its dependencies, Etc.

'Being always desirous of promoting the public good, we do join in the same sentiment with the committee, by acceding to their propositions in the manner following.

By the present, I do hereby accede to the four foregoing propositions, established and agreed upon for the purpose of establishing the peace and tranquility of the country, and that it may be constant and notorious, I sign the present under the seal of my arms, and countersigned by the secretary of this government at Natchez, the 22nds day of June, 1797.'

(Signed) Manuel Gayoso de Lemos.
(Signed) Joseph Vidal, Secretary.

On the 23rd the Governor and his officers left the fort, and returned to their houses. Thus ended this formidable tumult, without a single act of violence having been committed by the inhabitants of the country, during the suspension of the government and laws, for the space of two weeks!

The propositions were immediately transmitted to the Baron de Carondelet for his approbation and signature, who agreed to them without hesitation, except to a part of the third; which relates to the transportation of any inhabitant out of the district as a prisoner. By a royal edict, trials for capital crimes must be held in New Orleans, and to capital crimes only the Baron's exception extended. It was out of his power to set aside a royal edict. This was well understood by the committee, when the propositions were under consideration; but it was thought best to require more than was expected, to obtain as much as was necessary for the safety of the people.

As the Governor retired from the fort on the 23rd, he called at my quarter where we had a conversation of considerable length upon the state of the country.

During this interview the necessity of electing another committee, to aid in preserving good order and the peace of the country, was strongly impressed upon the mind of the Governor, who appeared fully aware of the propriety of the measure, and the day following issued his proclamation for that purpose.

The election took place about the beginning of July, when the following gentlemen were chosen. Joseph Bernard, Peter B. Bruin, Daniel Clark, Sr.,

Gabriel Benoist, Philander Smith, Isaac Gaillard, Roger Dixon, William Ratliff and Frederick Kimball.

The election of this committee, as was really intended on my part, put the finishing stroke to the Spanish authority, and jurisdiction in the district. The members, with the single exception of Frederick Kimball whose sentiments were doubtful, and who fell below the line, were decidedly republican, and firmly attached to the government and interest of the United States. This committee held their first meeting about the 15th of July, at the house I occupied: all their subsequent meetings were held at the same place, although they were offered the use of government house, they declined to accept it.

Having been very minute, and I fear tedious, in detailing the difficulties which occurred in carrying the treaty into effect, in the account of the commotion, I shall in the next chapter take up a subject more interesting.

CHAPTER IV.

Containing some account of the Mississippi river—of the settlements, and part of the adjacent country, Etc.

To say anything new respecting this river, whose magnitude and importance has for more than century, past, employed the pens of some of the ablest historians, philosophers and geographers of most nations in Europe, as well as in our own country, is not to be expected from me. In following such characters I shall proceed with diffidence, and confine my remarks to that part of this celebrated river, which I had an opportunity of examining myself, and which lies between the mouth of the Ohio and the Gulf of Mexico.

The confluence of the Ohio and Mississippi rivers, is in 37°0′23″ N. latitude, and about $5^h55'22''$ 8W. from Greenwich, agreeably to the observations in the Appendix to this work: but since those observations were printed off, my friend Don Jon Joaquin de Ferrer, whom I have already had occasion to mention, has determined at my request, by one of Arnold's best chronometers in descending the river from Pittsburgh, to New Orleans, the difference of meridians between the mouth of the Ohio, Natchez, and the two places first mentioned; and by careful comparison of the difference of those meridians, the longitude of the mouth of the Ohio appears to be $5^h56'24''$, which exceeds the determination in the Appendix by 1'1″2: and by combining this determination with the observations in the Appendix, the longitude of the mouth of the Ohio will be $5^h55'38''.1$ west from the royal observatory at Greenwich, and which is the position I have given it in the map. I am well aware, that this position differs nearly two degrees in longitude, and 14 minutes in latitude, from our best maps or charts. I have not adopted this alteration without some hesitation, and should still have been more cautious, if I could have found any other authority in favour of the former position, than charts unaccompanied by my observations.

Those who are descending the Ohio and Mississippi, and have been pleased with the prospect of large rivers rushing together among hills and mountains, will anticipate the pleasure of viewing the conflux of those stupendous waters. But their expectations will not be realized, the prospect is neither grand nor romantic: here are no hills to variegate the scene, nor mountains from whose summits the meandering of the waters may be traced, nor chasms through which they have forced their way. The prospect is no more than the meeting of waters of the same width, along the sounds, on our low southern coast.

The great rivers after draining a vast extent of mountainous and hilly country, join their waters in the swamp through which the Mississippi passes into the Gulf of Mexico. This swamp extends from the high lands in the United States, to the high lands in Louisiana: through various parts of which the river has at different periods made its way. From the best information I could obtain, the swamp is from thirty-six to forty-five miles wide, from the boundary many miles up, (the whole of which is several feet underwater every annual inundation) and much the greater part of it lies on the west side of the present bed of the river. From the mouth of the Ohio, to the southern boundary of the United States, the Mississippi touches but two or three places on the west side that are not annually inundated, and even these are for a time insulated; but on the east side it washes the high land in eleven places.

The swamp appears to be composed of the mud and sand, carried by Mad river into the Missouri, and the Missouri into the Mississippi, to which may be added, the washing of the country drained by the Mississippi and Ohio rivers, with their numerous branches, which furnish a fresh stratum every inundation. This stratum is deposited upon a stratum of leaves and other dead vegetables; which had fallen the preceding autumn. These strata may be readily examined in many parts of the swamp, and banks of the river. The thickness of the deposited strata differs considerably, and principally depends upon the duration of the different inundations. In 1797, the inundation was complete by the last of February, and the river was not entirely within its banks till the beginning of September following. But in the year 1798, the inundation was not complete till after the middle of May, and the river was generally within its banks by the first of August. The mean perpendicular height to which the river rises, (at the time of the inundation,) above the low water mark at the town of Natchez, is about 55 feet.

In descending the river, you meet with but little variety; a few of the sand bars and islands, will give a sample of the whole. When the water is low, you have high muddy banks, quick-sands, and sand bars; and when full, you might almost as well be at sea: for days together you will float without meeting with anything like soil in the river, and at the same time be environed by an uninhabitable, and almost impenetrable wilderness.

This river, like all others passing through flat countries, and not checked or confined by hills or mountains, is very crooked as may be seen by the chart. This arises from a very natural cause, and may be explained in the following manner. Suppose in the figure below, the lines a b, and c d, to be the banks or margins of a portion of the river, and water moving the direction e f, but meeting with an obstruction at f, it will be reflected in the direction f g, and at g, as well as at f, the bank will be worn away. About h, an eddy will be formed, where sand, earth, and rubbish will be deposited, and continually increase the convex parts, while the concave parts will be worn away, and in time a loop will be formed something like the dotted curve line I, h, m, and n in the figure, which will increase the magnitude till the river aided by an inundation breaks through a shorter way, and the convex part will become an island. If the loop should be very large, and the water cease to have much current along it, the two ends in a short time will be filled up by the great quantity of mud and sand, which are constantly mixed with the water of the Mississippi, and a lake will be formed. These lakes are to be met with in various parts of the swamp, and evident marks of having been at some former period, a portion of the main bed of the river.

In consequence of the great body of water in the Mississippi, and the light and loose nature of the soil, the concave banks of the river are falling in, more or less during every general fall or rise of the water; and I believe but few people have ever descended it, in either of those states, who have not heard or seen large portions of the banks give way, which are instantly carried off by the current, and the earth, sand, and some of the rubbish again deposited in the eddies formed by the convex points below.

From what has been said, no general caution must necessarily present itself to those concerned in navigating the Mississippi, which is to avoid the concave banks or shores. Many fatal accidents have happened on this river, either through ignorance of the danger, or inattention to coming to at improper places on the shore to cook, procure fewel [fuel], or for other purposes.

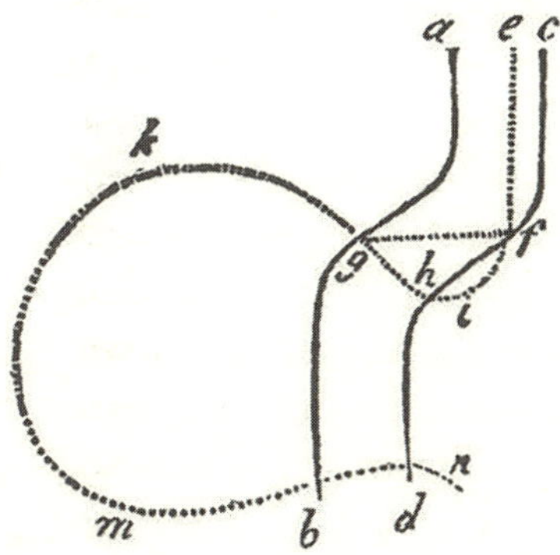

We have a late instance of a Mr. Mc'Farling, and part of his crew, being lost by the falling of a bank. When the banks are inundated they are less dangerous, being in some measure supported by the water, and not liable to give way; but the concave shores are still to be avoided, because the water near the bank and elevated above it, not being confined to the course of the river by the lower current, rushes straight forward among the cane and timber, and if Kentucky boats (as they are called,) fall within the draught of this upper current, it will be extremely difficult to relieve them, or prevent their being lost in the woods; many losses have been sustained from this cause.

A boat may at all times come to with safety at a sand bar, the upper or lower end of an island where young bushes are growing, or just at the beginning of [an] eddy, below any of the points that are covered with young cotton wood, (the populous deltoid of Marshal,) or willows (salix nigra.) From the mouth of the Ohio, down to the Walnut Hills, it is not safe to descend the river in the night, unless the boat be uncommonly strong, on account of the sawyers and planters. The former are trees slightly confined to the bottom by some of their roots or limbs, and the loose or floating ends continue a vibratory motion, generally up and down, some of them rise five or six feet above the water, every vibration. The latter are more dangerous, being firmly fixed or planted in the bottom. They are all easily avoided in daylight. With these precautions, the Mississippi may be navigated with as much, if not more safety, than any other river upon this continent.

It may generally be observed, that the banks of all our rivers subject to inundation, are higher on the margins of the rivers than some distance from them, and commonly terminate by a gentle declivity in a swamp: this is the case of the whole length of the Mississippi from the mouth of the Ohio, to the gulf of Mexico: hence some superficial observers, have been led to believe, that the river passes along the top of a hill with a valley on each side of it. After the water, at the time of the inundation rises above the banks, it runs from the river into the swamps with great rapid rapidity till they are filled to a level with the main current. Those swamps which communicate immediately with the gulf of Mexico, or the salt water lakes, never fill during the inundation; consequently the current continues over the banks into them till the waters fall.

Advantage hath been taken of this circumstance, in a number of places about New Orleans, for the erection of saw mills, which are found to answer a valuable purpose, and are kept constantly going during the whole term of the inundation.

The first large body of water which leaves the Mississippi, and falls by a regular, and separate channel into the gulf of Mexico, is the Chafalia [Atchafalaya]. It leaves the Mississippi in the westernmost part of that remarkable bend just below the boundary, and has every appearance of having been formerly a continuation of the Red river, when the Mississippi washed the high land from Clarksville, to the Bayou Tunica, (or Willing's creek,) the traces of which are yet visible by the lakes, through which a large current yet passes when the river is high. The distance on a straight line from Clarksville to the Bayou Tunica is not more than eight miles, but by the present course of the river, it is supposed not to be less than fifty miles. Should the Mississippi break its way through by a shorter course, (which is more than a mere probability,) the Chafalia will again become a part of the Red river.

When the Mississippi is high, the draught into Chafalia is very strong, and has frequently carried rafts, and likewise some few flats, or Kentucky boats down it, which are generally lost. This branch notwithstanding its magnitude is not navigable to the gulf of Mexico, owing to an immense floating bridge, or raft across it, of many leagues in length, and so firm and compact in some places, that cattle and horse are driven over on it. The surprising floating bridge or raft, is constantly augmented by the trees and rubbish, which the Chafalia draws out of the Mississippi.

During the annual inundation, a considerable stream of water called Bayou Manshack, or Iberville, leaves the Mississippi at Manshack, and after joining the Amit [Amite], falls into lake Maurepas, then into lake Pontchartrain, and communicates with the gulf of Mexico, near the mouth of Pearl or Half-

Way river. During the passage of the water along this channel, New Orleans stands upon an island, which may be considered the Delta of the Mississippi. This channel might be rendered navigable for boats during the inundation by removing the timber and rubbish, with which it is at present choked up: and no doubt, when the population of the country will justify the measure, it will be made sufficiently deep to communicate with the Lakes Maurepas and Pontchartrain, and the coast of West Florida at all times, which will be found a much easier and cheaper conveyance than by the way of New Orleans.

Below Manshack a few miles, but on the west side of the river, another branch called the Plaquemin, makes out of the Mississippi, and communicates by several mouths with the gulf of Mexico: from this branch there is a water communication with the Opelousas.

Between Evan Jones's place and the church, another branch makes out on the west side of the river, called La Fourche, (or the Fork,) from its dividing and falling into the gulf of Mexico by two branches. From the Mississippi along this branch to the sea coast, the settlement is compact.

Between New Orleans and the Balise [a fort and pilot station at the mouth of the river], the Mississippi has several communications with the gulf of Mexico; but they are generally too shoal to be of much importance.

After this short account of the river, I shall proceed to take some notice of the settlements on it below the mouth of the Ohio.

Ferdinand de Soto, after traversing the Floridas, discovered the Mississippi in the year 1541. The place where he first struck it is uncertain, but generally supposed to have been about the latitude 34°10′ north.

Notwithstanding this early discovery the river remained unexplored till 1682, when the enterprising [Rene-Robert Cavelier, sieur de] La Salle, in an excursion from Canada fell upon it, and followed its courses to its mouth; and is said to be the first person, who discovered that it discharged its waters into the gulf of Mexico. He was afterwards treacherously murdered on the coast, about one hundred leagues west of the Mississippi by his own men, where he was endeavouring to establish a colony.

No permanent settlement was attempted on the Mississippi till about the year 1712, when [Antoine] Crozet obtained a patent for exclusive trade of Louisiana: but previously to the obtaining this patent, a few miserable families had settled at, or near the place where New Orleans now stands. These settlers were the remains of two small colonies, that had been settled on the coast, east of the Mississippi in the years 1699, and 1701. One of these colonies was settled at the bay of St. Louis, and the other near the bay of Mobile. But the country being low, sandy, and miserably poor, and producing little beside pitch pine, added to the coast being too shoal for any other than the coasting trade, (of which at that time there was none,) those colonies after enduring innumerable hardships, and a number of the adventurers dying for want of the necessaries of life, the remainder quitted their habitations, and followed the coast to lake Pontchartrain, which they likewise coasted to the Bayou St. John's , and after ascending it few miles they crossed to the banks of the Mississippi, the distance across being less than a league.

Here those miserable people found a soil, not exceeded by anything in the world for fertility; but the banks of the river being subject to the annual inundation, their situation was still deplorable. Immediately after Crozet obtained his patent, a number of adventurers, (some of whom were very respectable,) arrived, who began with spirit to bank out the river, but the labour being great, the progress was small, until Crozet's patent was relinquished to the Mississippi company in the year 1717. The plan of this company [Company of the West] was projected by John Law, a Scotsman, during the regency of the Duke of Orleans, and shared the fate of all speculations which have paper only for their foundation.[23] The failure of this celebrated company checked the progress of the settlement; but it had taken too firm a hold to be abandoned; the banking out of the river had been carried on with great spirit, and the plantations thus protected, were not exceeded by any thing in America for beauty and fertility. Notwithstanding the misfortunes which had attended the first adventurers, the settlement extended as early as 1727, up to the present town of Natchez, where a fort was erected, but owing it is said to the improper conduct and tyranny of the French, the Indians to get rid of their troublesome guests, in a treacherous and barbarous manner, about the year 1731 massacred almost every French inhabitant in the upper settlement: but the year following, the French retaliated in a manner equally treacherous and nearly extirpated the whole nation of the Natchees; the few that escaped, took refuge among the Chocktaws, and other neighbouring nations, where some of their descendants yet remain.[24] From this time the

23. For an overview of the schemes of John Law, see John G. Clark, *New Orleans 1718–1812: An Economic History* (Baton Rouge: Louisiana State University Press, 1970), 9, 17–18, 22.

24. The history of French settlement among the Natchez Indians is described by Antoine Simon le Page du Pratz (ca. 1695–1775), who wrote the three-volume *History of Louisiana* (1758).

French had possession of the country until the peace of 1763, when all the French and Spanish possessions east of the Mississippi, except the island of Orleans, which has already been designated as the Delta of the Mississippi, were ceded to the crown of Great Britain. A few years afterwards, by an agreement between the crowns of France and Spain, the latter became possessed of Louisiana and the island of Orleans, to which they yet belong.[25]

The country on the east side of the Mississippi, and north of the island of Orleans, was peaceably possessed from the peace of 1763, till after the commencement of our revolutionary war in the year 1775, by Great Britain. In the mean time, the country increased rapidly in population, by the number of adventurers that went into it from England, Ireland, Scotland, and the colonies, now the United States. Among those adventurers was the late Mr. James Willing of Philadelphia, who resided on the Mississippi several years, during which time he became acquainted with the inhabitants generally, and their particular sentiments, as they relocated to the contest between Great Britain and her colonies. After the commencement of hostilities, he returned to Philadelphia, and was entrusted with a command down the Mississippi. On his arrival at Natchez, he entered into an agreement with the inhabitants, by which they were permitted to enjoy a state of neutrality during the controversy. As Mr. Willing conceived Mr. Anthony Hutchins to be the most dangerous and determined royalist in the settlement, he took him a prisoner to New Orleans; from whence he made his escape, returned to Natchez, and put an end to the neutrality. The particulars of this important possession of that country at least fourteen years, will be found in the following deposition, which is only one, out of several others to the same import, sent to the department of state in year 1797. [The lengthy deposition of November 6, 1797, not included here, was signed by Captain William Ferguson, in the service of the United States, and describes events of 1778 on local affairs in Natchez during that year regarding American or British loyalties.]

Thus ended the neutrality after a continuance of a few weeks. The settlement [Natchez] now became again subject to his Britannic Majesty, under the auspices of Col. Hutchins, in which state it continued till the 21st of Sep-

25. Ellicott is mistaken here as to the time frame. Louisiana was ceded to the Spanish on November 3, 1762, by France, her ally in the war against Great Britain—a year before the treaty in 1763 ending the French and Indian War, as it was called in the colonies, or the Seven Years' War, as it was known in Europe. Spain's decision to accept Louisiana was based upon her need to protect Mexico from an expanding British empire in North America.

tember 1779, when it was surrendered by Col. Dickson, who commanded the British forces at Batton [Baton] Rouge, to General [Don Bernardo] Bernard de Galvez, who commanded the troops of his Catholic Majesty.[26] From this time until after my arrival in that country, subject to Spain, except for a few weeks during an insurrection in the year 1781, in which Col. Hutchins was a principal actor.

From Natchez up to the mouth of the Ohio, there are no settlements of much account. Those at the Walnut Hills and Chickasaw Bluffs, consist of but few inhabitants, and they are of that class, who support themselves by supplying the soldiers with vegetables and other articles they may be in want of, and may be more properly denominated suttlers than settlers. The Spaniards have a small settlement at New Madrid, which was planned by Mr. [Diego de] Gardauqui his Catholic Majesty's minister to the United States. Col. [George] Morgan of New Jersey was some time Governor of that colony which was evidently on the decline when I descended the river.

The soil in the district of Natchez is at present uncommonly fertile: but the country being high, hilly and broken, the fine rich mould [topsoil] will be washed away after a series of cultivation, and the country become less productive. This tract of good upland is not very extensive, being about 130 miles in length along the Mississippi, and not more than 23 in breadth.

The staple commodity of the settlement of Natchez is cotton, which the country produces in great abundance, and of a good quality. The making of indigo, and raising tobacco, were carried on with spirit some years ago; but they have both given way to the cultivation of cotton. The country produces maize, or Indian corn, equal, if not superior to any part of the United States: the time of planting it is from the beginning of March, until the beginning of July. The cotton is generally planted in the latter end of February, and the beginning of March. Rye has been attempted in some places, and raised with success; but wheat has not yet succeeded. Apples and cherries are scarce, but peaches, plumbs and figs are very abundant. The vegetables of the middle states generally succeed there. The sugar cane has been attempted in the southern part of the district, near the boundary; I have not yet heard with

26. On the Gulf Coast conquests by Spain in British Florida, see John Walton Caughey, *Bernard de Galvez in Louisiana, 1776–1783* (Berkeley: University of California Press, 1934); Coker and Rea, eds., *Anglo-Spanish Confrontations;* John Francis McDermott, *The Spanish in the Mississippi Valley* (Urbana: University of Illinois Press, 1974); and Ralph Lee Woodward, ed., *Tribute to Don Bernardo de Galvez* (Baton Rouge: Moran Industries, 1979).

what success: but from Point Coupie [Coupee], down to the gulf of Mexico, it answers at present better than any other articles; and sugar has within a few years past become a staple commodity of that part of the Mississippi. A variety of oranges, both sweet and sour, with the lemons, are in great plenty on that part of the river.

From the great number of artificial mounds of earth to be seen through the whole settlement of Natchez, it must at some former period have been well populated. These mounds or tumili, are generally square and flat on the top: add to this circumstance in favour of the former population of that district, the following fact, which is very conclusive. In all parts where new plantations are opened, broken Indian earthen-ware is to be met with, some of the pieces are in tolerable preservation, and retain distinctly the original ornaments; but none of it appears to have ever been glazed.

The climate is very changeable during the winter, as may be seen in the Appendix; but the summer is regularly hot. During my residence at Natchez the greatest degree of cold was about 17° and that of heat 96° by Farenheit's scale.

The permanent degree of heat may be stated at about 14° beyond that of Pennsylvania. The conclusion is drawn from the following facts. In Pennsylvania the mean temperature of the best spring and well water, is about 51° and from the Mississippi, east to the Atlantic, in the parallel of 31° I found it about 65°, the difference is 14°.

The natives of the southern part of the Mississippi are generally a sprightly people, and appear to have natural turn for mechanics, painting, music, and the polite accomplishments, but their system of education is so extremely defective, that little real science is to be met with among them. Many of the planters are industrious, and enjoy life not only in plenty but affluence, and generally possess the virtue of hospitality, which never fails to impress the stranger and traveler with a favourable opinion of the country, and its inhabitants.

A large proportion of the inhabitants below the boundary, are descended from the original French settlers, and adventurers, and still retain in a great degree, the language, manners and customs of their ancestors.

The horses are tolerably good, but not equal to those of Pennsylvania. Many of them have been taken wild on the west side of the Mississippi, and are used both for saddle and draught. Few animals are reduced to obedience, and taught subordination more readily.

Contrary to my expectation, I found the cattle in the settlement of Natchez but little inferior in size to those of the middle states. They are extremely numerous, and it is not uncommon for the wealthy planters to possess from one to two hundred head, and sometimes more. The cows yield much less, and poorer milk than those of the northern states.

The domestic animals of that country, are not as gentle as those of the northern states. This probably arises from the little dependence they have upon man for their support. In the middle and northern states, they have to look up to their owners for their daily support: but in that country they can pass through the winter without his aid. If this idea be correct, may not the domestic situation of animals be considered a forced state and will not the degree of their domestication be in proportion to their dependence.

The mutton of that country is well tasted; but the wool of the sheep is more hairy, and less valuable than in the middle and northern states.

The hogs are but little, if anything inferior to those of any part of the United States, provided they receive the same attention.

The wild animals of the middle states will generally be found as far south as the Boundary, except the red fox, (canis vulpes,) ground hog, (arctomys monas,) and musk rat, (mus zibethicus,) but they are evidently somewhat smaller.

I shall now proceed to give some account of the construction of the map of the Mississippi. A continued, and correct survey of this river will scarcely ever be obtained on account of the swamps, lagoons, thickets, and cane brakes on its banks: and below the banks, the impediments will be equally great. In some places impassable quick-sands will be met with; in others the water will be found washing the high, and almost perpendicular banks, and no place left for foothold. Some other mode, different from the common method of surveying must therefore be resorted to. The following was used in constructing the map to which this refers. The mouth of the Ohio and town of Natchez, were taken as given points, both as to latitude and longitude. An excellent surveying compass, corrected for the variation of the needle, was used in taking the courses, which were entered in time, instead of space. Every day in descending the river when the sun shone at noon, this meridional altitude was taken, from the artificial, or reflected horizon, with an excellent sextant made by Ramsden, which graduated by vernier to twenty seconds, and was generally found by a very great number of observations to determine the latitude true, within less than a minute. The latitudes determined by those observations, are entered on the chart of the river at the places where the observations were made: all the courses between each two of those points were protracted in time, instead of space, that is by calling the time space. Each set of courses were then expanded, or contracted, so as to agree with the points of latitude to which they belonged. From the number of latitudes taken, I expect that no part of the river will be found erroneous in that respect: so much cannot be said in favor of the longitude, except at the mouth of the Ohio, and the town of Natchez, which are considered as given points: the latitude and longitude of the latter being determined with as much precision as any point within the United States.

Since my first chart of the river was made for the use of our government, I have been enabled to correct some points of longitude between the mouth of the Ohio and Natchez, from the observations of Mr. Ferrer.

In consequence of the banks of the river constantly giving way, no map or chart of it can be expected to be tolerably correct for more than a century to come. The chart of the Mississippi is divided into two parts, for the same reason already given for the division of the Ohio: the division takes place at latitude 33° north.

CHAPTER V.

Containing an account of some occurrences which took place in the settlement of Natchez after the election of the second, (commonly called the permanent,) committee, until the author left the Town to commence the demarcation of the boundary.

The occurrences which took place in the settlement of Natchez after the termination of the commotion, and the election of the permanent committee, until my beginning the demarcation of the boundary, must daily become less interesting; they will therefore generally be but slightly mentioned, and some of them wholly omitted. The difficulties were however as great, if not greater, than any we met with from the Spanish government, but as some of them arose from the conduct of certain citizens of the United States, a few of whom did at that time, and still continue to enjoy the patronage of our government, it is not impossible but they acted in conformity to directions they had received, or agreeably to a plan previously devised. It would therefore be improper to bring their conduct before the public, which possibly had its origin from a higher source, and which was fully competent to decide on its propriety.

I have ever been an enemy to the abuse of the officers of government, unless after conviction by a competent tribunal; it is lessening them in the esteem of the public, and not infrequently renders both a useful office and officer contemptible: added to which, it is either insulting the judgment, or arraigning the motives of the executive, who made the appointments.

The press cannot be too free, when directed to a manly, and candid discussion of principles; but its usefulness ceases, when it waves the principle, and attacks the individual. From a full persuasion of the correctness of this opinion, in opposing plans brought forward, or seconded by the officers of the United States, which appeared injurious to the interest of my country, the plans, and not the individuals, (when they could possibly be separated,) have felt my opposition.

After Governor Gayoso had issued his proclamation for the election of the permanent committee, a number of gentlemen were anxious to have Mr. [Anthony] Hutchins elected a member of that body; not from any influence he had with the well informed part of the settlement, but from his being popular with the class of the populace, who regard men more than principles. Contrary to expectation, he declined serving, on account (as he said) of his age and infirmities.

The committee was no sooner organized, than it was evident its measures would be directed to the attainment of two objects: *first,* the securing the country to the United States, and, *secondly,* the preservation of peace and good order in the settlement. The first was contrary to the wishes of the officers of his Catholic Majesty, and the second to those attached to the British interest, to which may be added another class, who had nothing to lose, but hoped to gain by the tumult and disorder.[27]

Mr. Hutchins attended the first meeting of the committee as a spectator, but finding the views of the body, so different from his own, he became discontented with its proceedings, which circumstance was immediately taken advantage of by Governor Gayoso, who expected by dividing the inhabitants, to regain the power he had so lately lost.

On the 26th of July, Governor Gayoso received his appointment from the Court of Madrid as Governor General [of Louisiana and West Florida] in place of the Baron de Carondelet, who was promoted to the government [presidency] of Quito. On the 27th, Governor Gayoso informed me by a polite note of his promotion, and that he should leave the settlement on the 30th, and proceed to New Orleans; that for the present Captain Minor would

27. Colonel Hutchins simply disapproved of any leadership other than his own, and what he regarded as the "British" interests; he opposed both the Spanish officers and those who favored the United States, and/or their own selfish interests. As to the factions active in Natchez: "In one group were 'those who complained that Governor Gayoso had, at various times, insulted, wronged, and refused to grant land to them.' They included Colonel Anthony Hutchins, the vitriolic Tory whose extensive landholdings and irascible temper tyrannized his acquaintances and family alike" (qtd. in Holmes, *Gayoso,* 187; on Hutchins, see also ibid., 191, 257). Ellicott later (1799) acquired evidence of Hutchins's position as a major in the British army.

represent him in the government of Natchez. But previous to his departure, he directed Captain Minor not to oppose Mr. Hutchins in the election of another committee, if he should attempt it.

A few days after the Governor's departure for New Orleans, Mr. Hutchins came to the house I occupied, and requested my aid in dissolving the permanent committee, which was then in session, and to let the principal power be lodged in his hands and that of another committee, which he would have elected. This proposition met with my decided disapprobation, not only from its being unnecessary, but from the impropriety of being instrumental in placing a British officer, in a situation, by which he might be enabled materially to injure the United States. Notwithstanding my opposition and remonstrances, Mr. Hutchins went into the hall where the committee was on business, and after a few preliminary observations accompanied with abusive language, he told the members they were 'no committee, that they were dissolved, and he would direct the election of another, that it was a hard pill, and rough burn, for them to swallow, but he would force it down their throats.' To which the honourable Judge [Pedro Bryan] Bruin (who was a distinguished officer in our revolutionary army,) coolly replied 'Col. Hutchins you appear to be acting the dictator, and at the same time affect to be waving the American cap of liberty; but they are incompatible, the cap of liberty but ill becomes you, who opposed in arms the independence of the United States.' Mr. Hutchins then left the hall exclaiming, 'you are no committee, you are dissolved!'

In this extraordinary proceeding Mr. Hutchins was seconded by some citizens of the United States, several of whom held commissions under our government. It is possible they were deceived.

Immediately after this singular transaction, Mr. Hutchins applied to Captain Minor, and obtained permission to have another committee elected: but this election met with great opposition, and was generally considered by the well disposed inhabitants, as an artful measure, calculated to divide the people between two committees, which if effected would, in all probability, end in a breach of the neutrality by one, or the other of the parties, and thereby produce the reestablishment of the Spanish government with all its terrors. Under this impression, six out of ten subdivisions of which the district was composed, protested against the election, of course there were but four elections holden [held] agreeably to the permission obtained from Captain Minor. These four returns were sent to Mr. Hutchins sealed up, who presented them to the judges, but they refused to act, he then presented them to Captain Minor who declined having anything further to do with the business.

On the day appointed for the meeting of this new committee, the four members elected by the four subdivisions, with Mr. Hutchins assembled at Mr. Belts to proceed to business: but the difficulty which first attended opening the returns still existed, however men void of principle, and determined to carry a favourite object, are seldom arrested in their progress by considerations of propriety. Mr. Hutchins therefore resolved to stop the first three persons who might be passing along the road, and constitute them judges of the returns of the election. A short time after he had taken this resolution, Mr. Justice King, (an equivocal character,) and his two sons Richard, (a Captain in the Spanish militia,) and Prospect, who were on their way to the town of Natchez, were called in by Mr. Hutchins, who constituted them judges, after which they opened the returns, and declared Thomas Green, James Stuart, a Mr. Ashly, and a Mr. Hocket, to be elected.

Mr. Green, (although a Captain in the Spanish militia,) I always found to be both a republican and friend to the United States; but being an easy, quiet man, and not calculated to fathom political intrigue, it was supposed that he would on all occasions be influenced by Mr. Hutchins, with whom there was a family connection. Mr. James Stuart was only entitled to notice from the active part taken by himself, and some of his relations in favor of the British during our revolutionary war.

Mr. Ashly was a Baptist minister, he went to Natchez from the state of Georgia, and it was affirmed by most respectable authority, that he was followed for taking property not his own.

Mr. Hocket's honesty I never heard called in question, but his other qualification were below mediocrity.

From this statement which is certainly correct, it appears that but four, out of the ten members were duly elected. The elections were holden on Saturday the 2nd of September 1797, and three days after, the four members already mentioned, with Mr. Hutchins at their head, proceeded to business: on the 9th following, an itinerant attorney, of some education abilities, by the name Shaw, held an election, (without any other authority than from Mr. Hutchins,) in the subdivision of Homochitto; when a Mr. Davis was elected by ten voices, no return was ever made of this election, but Mr. Davis nevertheless took his seat. He was justly esteemed a steady friend to the United States, and being independent in his sentiments, he disapproved of great part of the proceedings of Mr. Hutchins's committee: The week following, Mr. Dayton who I have already had occasion to mention, procured the election of Mr. Justice King and Mr. Abner Green, son-in-law to Mr. Hutchins, by subscription: the number of signers did not exceed thirty. This method was resorted to, because the inhabitants of the two districts for which these men were returned, had protested against the election of Mr. Hutchins's committee: this private mode by subscription was therefore resorted to, as a general election could not be had. The two last members appointed, served to increase the number of the committee, which yet consisted of but seven instead of ten members.

This imperfect representation of the settlement under the direction of Col. Hutchins, after a session of three weeks, produced a petition and memorial to congress: which was equally remarkable for its misrepresentation, composition, and absurdity of its requests or prayers, for it contained many; but its great length would not justify the insertion of it in a work of this kind.

One of the most prominent features of the petition, was a paragraph calculated to injure Lieut. Pope and myself with our country, and at the same time to furnish the minister of his Catholic Majesty with materials to enable him to make good his charges against us. This petition and memorial, after being finished by Mr. Hutchins, and agreed to by the committee, was circulated with great secrecy, and only submitted to the perusal of Mr. Hutchins's confidential friends, and when read at small, partial meetings, for the purpose of obtaining signatures, the paragraph reflecting upon 'the persons sent there by the United States,' (meaning Lieut. Pope and myself,) was generally, if not always omitted. It was by accident that a true copy of it was obtained. Mr. Hutchins had sent one on, with a number of papers, by way of New Orleans, for the Spanish Minister, which were intercepted, opened, and handed to me. The Petition and memorial was immediately copied by the late Mr. Benoist, chairman of the permanent committee, and my secretary Mr. Gillespie, after which the packet was made up and forwarded. From this circumstance I was enabled to furnish the department of state with a copy of the petition and memorial, accompanied with remarks, some time before the original forwarded by Mr. Hutchins and committee, was received.

Upon shewing a copy of the petition and memorial, to some of the signers, they were astonished at the deception which had been practiced upon them by Mr. Hutchins and his committee, and declared in the most solemn manner that part of it had never been read to them, in consequence of which, a number of protests were drawn up and signed, by many of the respectable inhabitants against the paragraph above alluded to, and presented to me for the purpose of being forwarded to the department of state, to counteract any impression that might otherwise have been made by the deception. These protests were all of the same import, the following is an exact copy of one of them.

> 'Whereas there has been a memorial drawn up by Col. Hutchins, (and a committee stiling [styling] themselves a committee of safety and correspondence,) to the congress of the United States, intended as we conceived to represent only the grievances of the people of this government; but among other representations, we find since signing it, a paragraph reflecting very severely upon the conduct of the honourable Andrew Ellicott, Esq. and Lieutenant Pope, commanding the troops of the United States here, for exciting the inhabitants of this place to an insurrection[against] the government.
>
> Therefore we the subscribers, as conceiving ourselves over-reached in this particular, not comprehending the meaning of such paragraph, or as we are more inclined to believe, not hearing it read by the person, or agent employed for that purpose, and being now willing to do justice to the characters of the said honourable Andrew Ellicott, Esq. and Lieut. Pope, do most solemnly protest against such paragraph, and make this our recantation, at Natchez, October 24th, 1797.'
>
> (Signed by) Baley Chancy, John Erwin, and many others.

Mr. Ashly, Baptist minister, whose name has already been frequently mentioned, with a number of his friends who had entered their names on the petition and memorial, (as it was called,) finding that it contained a paragraph which was not read to them when they signed it, waited upon Mr. Hutchins and requested their names to be stricken out, unless the obnoxious paragraph was expunged. To which he replied 'I am much less anxious about retaining any other part of the memorial, but I now have your names, and shall make what use of them as I see proper.'

After this transaction, Mr. Hannah immediately called upon me, much alarmed on account of Lieutenant Pope and myself, as it was now evident our injury, more than the good of the settlement, was Mr. Hutchins's real object, and proposed several plans to obtain and destroy the petition and memorial;

these plans were not only objected to as improper; but Mr. Hannah was requested to oppose every attempt to get possession of it for any such purpose. He was told that no injury could arise from it, because the government of the United States was not a government of mere opinion, in which an individual could be affected by the caprice of any man, whatever might be his talents, and standing in the community; but a government of laws which served equally for the direction of the first magistrate and the private citizen. That the executive would be cautious in censuring or dismissing a person, (who had devoted more than twenty years of his life in various employments to the service of his country,) without either a hearing, or an investigation to gratify any man, or party of men, much less a stranger almost unknown, but for the part he had taken in opposition to the independence of the United States, and in favour of monarchy.

The declaration of Mr. Hutchins relative to that particular paragraph of the memorial being made public by Mr. Hannah and his friends, turned the popular torrent against him so effectually, that his influence was but trifling during my subsequent continuance in the settlement, though he frequently endeavoured to regain it by artifice and slander.

Notwithstanding the opposition I had uniformly given to taking the memorial, or arresting it on its way to the seat of our government, the irritation was so great that the packet which contained it, together with many other papers, was intercepted and brought to me unopened, in which condition it was returned to Mr. Hutchins, and his committee, who to the best of my recollection met afterwards.

Having now given an account of the election of Mr. Hutchins's committee, and the sanction it gave to his labours, every part of which can be substantiated by documents furnished to the department of state where they no doubt remain filed, I shall return to the history of the proceedings of the permanent committee, and of my own conduct, which were too intimately connected to be wholly separated. Notwithstanding the rude, and almost unprecedented manner in which Mr. Hutchins treated the permanent committee, and which was seconded by menaces from various parts of the country, that body continued firm in the discharge of its duty.

In the beginning of September 1797, I received from the department of state a communication, containing an account of the detection of Mr. Blount's plans [see note 19] for conquering Louisiana and the Floridas, for his Britannic Majesty; which was the second dispatch I had received since I left the seat of our government, the other received on the 25th of July preceding. The communication relative to Mr. Blount was immediately laid before the permanent committee, and measures taken to prevent any part of the plan being carried into effect, if attempted. The traders and interpreters among the Indians were written to upon that subject, and requested to be vigilant in preventing the circulation and suppression of any improper talks that might be sent among them.

From the part which Mr. Hutchins took in quieting the commotion, it was my opinion for some time after I had received the communication, that he was unacquainted with Mr. Blount's designs, notwithstanding his interviews with one of his friends who had left Philadelphia the preceding December, as stated in the third chapter; but his subsequent conduct compelled me to change this opinion, and to conclude that his opposition to the commotion arose from its being either premature, or not in unison with Mr. Blount's plans. For it is to be observed, that the country had scarcely become tranquil, before Mr. Hutchins began to form a party against the permanent committee, and was desirous of having the principal power placed in his own hands; to accomplish which no art was left unpractised and no falsehood but was propagated to secure this favourite object. Could this be done to serve the United States? Certainly not; he was well known to have been at all times an enemy to them, and to republican forms of government; and was moreover at that very time an officer on the British military establishment. It is likewise a fact, that three days previous to my arrival at Natchez, when it was known that I was on the river, he went and renewed his oath of allegiance to his Catholic Majesty. In consequence of his inflammatory libels, which he industriously circulated through the settlement, he brought many well meaning inhabitants into the belief that they were in a state of slavery, (though at that time the most free people upon the continent,) and that nothing could prevent their chains being riveted, but a spirited opposition to the permanent committee, and 'some of the persons sent there by the United States.' Those misrepresentations had nearly proved fatal to the district by arming one part of the inhabitants against the other. The inhabitants of Bayou Piere [Pierre] being the most distant from the seat of information, were the most easily imposed upon, in consequence of which I received the following letter from the permanent committee.

Committee room, 12th September, 1797.

Sir,

Information has been received by a member present, that forty armed men were on their way from the Bayou Piere, and that the inhabitants of Coles-Creek were in an ill humour, and threatening to arm also. The designs of those preparations seem either to be to interrupt the permanent committee in its proceedings, or otherwise to act hostilely against it. The committee conceive it their duty to make this communication to you in consequence of your having made yourself guarantee for the preservation of peace and good order in this country, by the instrument of writing you was pleased to sign the 22nd of June last.

I have the honour to be, with great esteem,
Your most obedient humble servant,
(Signed) Joseph Bernard,
Chairman of the permanent committee.

Honourable Andrew
Ellicott.

To the foregoing letter, the following answer was immediately returned.

Natchez, September 12th, 1797.

Sir,

I am sorry to find from your communication of this date, that some of the inhabitants of Coles-Creek and Bayou Piere, continue their opposition to good order. You may depend upon my influence and exertions, both in supporting your authority, and preserving inviolate the present neutrality of this district.

I have the honour to be, with great esteem,
Your friend and humble servant,
Andrew Ellicott.

Joseph Bernard, Esq. chairman
Of the permanent committee.

The good sense of the people alluded to in the communication from the permanent committee, aided by proper advice, prevailed over their prejudices and passions, and they returned home without committing any act of violence.

On the 14th of September, I received the following letter from the permanent committee.

> Sir,
>
> The many proofs that the committee has had of your desire to contribute to the welfare of this country, encourage it to request of you the service mentioned in our first resolve of yesterday, of which this encloses a copy.
>
> I am, sir, your most obedient servant,
> (Signed) Joseph Bernard,
> Chairman.
>
> Committee room, Sept. 14th, 1797.
> Honourable Andrew
> Ellicott.

The resolution was in these words.

> Sept. 13th. *Resolved,* That the 5th resolve of the 29th ultimo be rescinded, and that Mr. Ellicott, whose inclination for the interest and happiness of the country, we have a convincing proof of in his former communications to the general government published since, and now in our hands, and who from his residence for several months among us, is well acquainted with the circumstances of this country, be requested to represent our present situation to his Excellency the President of the United States, and likewise all the measures which he shall deem conducive to the future welfare of this country, in the event of the late treaty between his Catholic Majesty and United States being carried fully into effect.
>
> (True copy.)
> (Signed) G. Benoist, Secretary.

The fifth resolve of the 29th ultimo, mentioned in the above resolution, was rescinded, as not being sufficiently comprehensive, and the object included in that of the 13th.

By the resolution of the 13th, the committee imposed a very difficult task upon me; but with which I nevertheless endeavoured to comply, and after my ideas were thrown together, they were submitted to that body, and approved of before they were transmitted to the department of state. As that communication contained my sentiments, (which have been extremely misrepresented,) relative to the local and political situation of the country, I shall give the substance of it from my original notes.

The *first,* and principal object with the committee was a constitution, or form of government, and though not expressly mentioned in the resolve, (on account of the neutrality, and the Spanish government having yet a nominal existence,) was nevertheless well understood.

On the subject of government, there have ever been such a variety of opinion, that there is no plan which human wisdom could devise that would not find opponents, and none so bad, that would not have advocates. This diversity of opinion arises in part from habits early acquired, and which vary gradually with the different stages of civilization, and unequal distribution of wealth in a country.

In framing a government for a district or settlement, it will be proper to inquire into the population, habits, state of civil society, and circumstances of the inhabitants: because these for a time, may be such that a government purely representative, would be very improper, by being placed completely in the power of the vicious, indigent, and uninformed, would oppress the wealthy and industrious, on whom the burden of supporting government generally falls; and which would probably be the case at present in Natchez, for this like other new countries, is peopled by the following descriptions of persons, viz. people of ambition and enterprize, who have calculated upon an increase of fame and fortune, others who have fled from their creditors, and some, (not a few,) from justice; to which may with propriety be added, those who fled from the United States during the revolutionary war, for their monarchical principles, or treasonable practices.[28]

When settlements as they grow up from the causes above mentioned, become constantly incorporated with old and stable governments, they generally produce but a small inconvenience by having a proportionable share in the general representation; but the effect would be very different, if such

28. Ellicott notes at this point: "This representation was made to the department of State, September 24th, 1797" ("Southern Boundary," vol. 1).

a settlement without any qualification of the electors as to property and allegiance, should be invested with the sovereign power of legislating for themselves. In case of such an event, creditors might expect to be injured, if not ruined, gentlemen of probity, worth, and information neglected, if not persecuted, and public confidence annihilated.[29]

It appears to me, that a territorial government similar to that of the North Western Territory, is less expensive, and better calculated than a representative one, for doing justice in a district populated from the causes above mentioned, and where the habits of the people have in part been formed under a despotism, and by whom the principles of representative government must be but imperfectly understood, and the free white population supposed not to exceed five or six thousand souls. The Governor, in conjunction with the judges, being competent to the selection and adoption of laws for the district, from the codes of the different states, (comprised in the Union) and those laws thus adopted, being again subject to the approbation of congress is as great change from despotism, towards representative government, as ought suddenly to be made in the situation of any people, (however enlightened,) until their habits, circumstances and morals become more congenial to the true principles of liberty otherwise there will be great danger of falling into licentiousness which is the natural extreme. This may be considered political heresy, but it will be found orthodox in practice.

Although domestic slavery is extremely disagreeable to the inhabitants of the eastern states, it will nevertheless be expedient to tolerate it in the district of Natchez, where that species of property is very common, and let it remain on the same footing as in the southern states, otherwise emigrants possessed of that kind of property, would be induced to settle in the Spanish territory. [Ellicott uses the legalistic term "property" to describe slaves at this point because Mr. Hutchins had represented him falsely as an ardent abolitionist.]

The manner of disposing of the vacant land, is a subject in which the inhabitants are materially interested. The mode heretofore pursued by the United States, would neither give satisfaction to the present inhabitants, nor in my opinion be good policy, setting aside the advantage it gives the wealthy, in a monopoly the most dangerous of any other to the liberties of the people. By disposing of the vacant land in small tracts, and at a moderate price, the

29. Ellicott is expressing here the Federalist view of mob rule, based upon their understanding of what happened during the French Revolution's "Reign of Terror" (1793–94), and the subsequent seizure of power by the Directory (1795–99).

preference being given to actual settlers, a firm, compact settlement would speedily be formed, which from its local situation, would be very advantageous to the United States in case of a war with Spain: another reason for this practice is, the danger of losing a number of our citizens, who would be induced to settle in the Spanish territory, where lands are obtained in any quantity, (great or small,) upon very easy and advantageous terms.[30]

There is yet one other source of uneasiness among the inhabitants, and which relates to their titles. It appears that much of the greater part of the lands now occupied, are covered by old British grants. The occupiers of those lands may be divided into two classes. *First,* those who continued in the country after its conquest by the Spaniards [1779–84], and renewed their titles under his Catholic Majesty, and *secondly,* those who are seated on old British grants, which became forfeited to the crown of Spain by their owners or attornies [attorneys], not appearing, and occupying them agreeable to the tenor of two proclamations or edicts, issued by his Catholic Majesty; the one dated in 1786, but whether this was the first or last, I am unable to say, as I have not yet been able to procure either of them.[31] The lands thus forfeited, have been granted by the officers of his Catholic Majesty, in the same manner, as practiced in granting vacant lands. This class of settlers may be considered as composing the body of the settlement. With respect to the first class, there cannot possibly by any doubt as to the validity of their titles: and the second, upon the principles of justice and equity, are perhaps equally safe; but they have their fears, and are therefore desirous that an act of congress may be passed confirming all their titles, that were good under the crown of Spain, at the time of the final ratification of the late treaty. So far the representation as made to the department of state, agreeably to the request of the permanent committee.

About the 20th of September, the committee met with a great loss in the death of its chairman, Mr. Bernard. He was a gentleman of good understand-

30. See Gilbert C. Din, "Spanish Immigration Policy in Louisiana and the American Penetration, 1792–1803," *Southwestern Historical Quarterly* 86 (1973): 255–76; and C. Richard Arena, "Land Settlement Policies and Practices in Spanish Louisiana," in *The Spanish in the Mississippi Valley,* ed. John Francis McDermott, 51–60 (Urbana: University of Illinois Press, 1974).

31. See the following studies on land grants: Holmes, *Gayoso,* 34–38; Edward F. Haas. ed., *Louisiana's Legal Heritage* (Pensacola. FL: Perdido Bay Press, 1983); John E. Harkins, "Legal and Judicial Aspects in Spanish Louisiana," in *Readings in Louisiana History,* ed. Glenn R. Conrad (New Orleans: Louisiana Historical Association, 1978).

ing, sound judgment, and of the most inflexible integrity. From his youth he was strongly attached to republican principles, and it may truly be said, that he expired serving the United States. His easy manners, benevolence and hospitality, were of that cast, that he only need be known, to be esteemed.

Upon the death of Mr. Bernard, Mr. [Gabriel] Benoist was unanimously elected chairman of the permanent committee, the duties of which he performed with singular assiduity, judgment and integrity. He was a French gentleman, of very respectable connections; but from an ardent passion for liberty, he left his own country, and espoused the cause of the United States, in their arduous struggle for independence; and afterwards retired into the settlement of Natchez, under the full persuasion, that it would be shortly annexed to the United States, where he married an amiable lady, the daughter of an honest, worthy planter, by the name of Dunbar.

Mr. Benoist was no sooner elected, than he was attacked with all the abuse, obloquy, detraction and misrepresentation, that could be devised by Mr. Hutchins and his party: his being both a Frenchman and republican, were sufficient causes to merit the hatred of Mr. Hutchins, by whom he was loaded with the epithets, French jacobin and democrat,[32] together with the whole catalogue of kindred terms, invented for party purposes, and intended by an association of ideas, to lessen the character of those whose conduct would not be injured by an exhibition of facts.

Mr. Benoist being a gentleman of great sensibility, could not but feel the injustice done to his character, and though his natural vivacity veiled from the public eye, the injury done to his feelings, it laid the foundation of a disorder, under which he sunk the following summer.

The permanent committee took little or no notice of Mr. Hutchins's committee; but depended on its own integrity, and conscientious conviction in its own mind, that its endeavours and exertions, were directed to the true interests of the United States for support.

32. The term "Jacobin" referred to the radical political party that controlled the French Revolution during the Reign of Terror (1793–95). Edmund C. Genet, French chargé d'affaires, during President Washington's administration, was linked to the Jacobins in France, and to fomenting pro-French disturbances in the West against the Federalist government, which led President Washington to demand his recall by the French government. Jacobins, therefore, were identified by legitimate governments everywhere as the party of revolutionary insurrection and global social disturbance. See Ernest R. Liljegren, "Jacobinism in Spanish Louisiana, 1782–1797," *Louisiana Historical Quarterly* 22 (January 1939): 47–97; and E. Wilson Lyon, "The Directory and the United States," *American Historical Review* 43 (1938): 516–17.

My communications to the department of state, and which I did not think myself justifiable in making public, (particularly that one drawn up at the request of the permanent committee,) were so wantonly misrepresented, it produced such a degree of irritation among some of the inhabitants, that I found it necessary for a short time to augment the number of my guard. To allay those prejudices, and if possible prevent the deluded from committing acts of violence, the following address was circulated through the district, and read by my secretary at a public meeting, about seven miles from the town.

> To the inhabitants of the district of Natchez.
>
> I have received certain information, that the following reports evidently calculated to injure me in the estimation of the good people of this district are in circulation. *First,* that I have in my official communications to the government of the United States, recommended that the vacant lands in this country, be laid off, and sold in squares of six miles to a tract. *Secondly,* that slavery ought to be prohibited here, as in the North Western Territory, and *thirdly,* that I am, or have been extensively concerned in large landed speculations, particularly in the purchase of old British grants.
>
> I now in the most unequivocal manner, declare the above reports to be without the least foundation. The first two, you will find contradicted by my official communications, when they are published. In those communications, I have recommended, that the vacant land be disposed of in such a manner, as to accommodate actual settlers of all descriptions, and that the system of slavery be continued upon the same footing, that it is in the southern states. The third charge, of being concerned in landed speculations, will in time be found equally false with the others.
>
> These reports have originated with the enemies of the United States, and are calculated to answer the worst of principles. They have been brought forward with so much art, and propagated with so much address, that many well disposed, honest, upright characters have given them credit. They are of the same stamp with Mr. Blunt's letter to [James] Carey [a fellow conspirator with Blount], and intended to weaken the confidence of the inhabitants of this district, in the agents of the United States, that in case of danger, they may have no rallying point, and thereby become an easy prey, to some of the European powers now at war.

My duty and inclination, have both impelled me to use all my interest with the executive of the United States in favour of this district, to which I feel a strong partiality, a partiality not of the moment, but the result of several months experience. These exertions in your favour will be continued with zeal and assiduity.

I am with sentiments of high respect,
Your humble servant,
Andrew Ellicott.

Natchez, October 13th, 1797.

Notwithstanding the industry of the designing and turbulent, in the propagation of reports injurious to the characters of the members of the permanent committee, as individuals, and to their conduct as a body, they in both capacities met with the most decided support from the honest, intelligent, independent, and well informed; who at all times have great weight, in all well regulated governments; and when they cease to be regarded by an executive, and lose their influence with the populace, either anarchy or tyranny is the certain consequence.

From the able support given to the committee, it was enabled to surmount all the difficulties, by which it was at various times surrounded, and triumph over the friends of disorder and confusion, who could not at that time on account of the neutrality, be publicly supported even by the Spanish government, whatever might have been its strength in political mercenaries, who are easily, in all countries converted into military ones.

Although the members of the committee felt themselves thus secure, they feared the effect of the misrepresentations of Mr. Hutchins, and his followers, upon the executive of the United States, so far as they related to Lieut. Pope, and myself. To counteract the effect of those misrepresentations, the committee drew up, and presented to me the following address, which though evidently intended to be made public, was only transmitted to the department of state.

To the honourable ANDREW ELLICOTT, commissioner of the United States of America, for running the line of demarcation between the United States and Spain,
Sir,

Permit the members of the permanent committee of the district of Natchez, to accost you with all the attachment and respect, which a just sense of your talents, patriotism, and truly amiable manners is calculated to inspire. To the judicious arrangements entered into with the Spanish government, and which you had the happiness to concert with the commanding officer of the United States troops, and with certain well disposed, and influential characters in this district, we feel ourselves indebted for the neutral position which we at present hold, and which it is our particular duty, interest, and inclination to preserve inviolate, by all the means in our power: And we consider it sir, as one of the most agreeable consequences of our neutrality, that it affords us the pleasure and advantage of a safe, and unreserved communication with you: it surely is but justice to add, that when difficulties and doubts, have arisen in the discharge of our duty, from the novelty of our situation, your friendly advice has contributed greatly to remove them, and that aided by your experience we have been enabled to act in a consistent, if not a popular part, although placed in circumstances embarrassing in the extreme, and altogether without precedent or example.

These, sir, are the sentiments of the committee, and it is with pain they observe, that through the restless and intriguing disposition of a designing individual [Hutchins], attempts have been made to injure you in the public mind, and to prejudice you, not only in this district, which you have so essentially served, but also with the executive of the United States, whose dignity and interest you have so well supported. This sir we consider a censure so ungenerous and unmerited, and the whole procedure so base and ungrateful, that so far from concurring therein, we should hold ourselves criminal if we remained silent on the occasion. With a view sir to do justice to your character, so far as it has been connected with the proceedings of this committee, and at the same time, to remove any improper impressions which the executive of the United States may have received, with regard to ourselves, from the misrepresentations of Col. Hutchins, we have thought proper sir, to offer you free access to our files with the privilege to make such extracts therefrom, as you will think necessary for your justification.

Although the committee have already made honourable mention of Lieutenant Pope, in one of their resolves, they cannot with justice close this address to you sir, without expressing the most unequivocal manner their approbation of his conduct in this business, and giving it as their opinion, he has deserved well of his country.

That your well meant endeavours to serve this district, may be productive of consequences salutary to the inhabitants, and honourable to yourself, is the sincere wish of

Sir, your very humble servants,
Gabriel Benoist Chairman

Philander Smith	Roger Dixon
Peter B. Bruin	William Ratliff
Daniel Clark	Frederick Kimball
Isaac Gaillard.	

Natchez, October 21st 1797.

The permission given in the foregoing address, to have access to the files of the committee, was never made use of; because it was impossible for me to suppose, that any measures injurious to the character, or reputation of a citizen, in a public capacity, would be attempted by the first magistrate of a great, free, and powerful nation, who held his office by the voice of a watchful people, (ready at all times to scrutinize his conduct, and make the cause of injured citizen their own,) without a regular investigation, to which every freeman ought to be entitled, and from which none but the guilty would shrink.

In November 1797, the appointment of Col. Grandprie [Carlos Louis Boucher de Grand-Pre] by the court of Madrid to the government of Natchez, and its dependencies was publicly announced. Immediately upon receiving this information, the permanent committee took the subject into consideration, and in a firm, and manly manner, resolved that Col. Grandprie (who was then at New Orleans) could not be received in the district of Natchez, in the quality of Governor. The proceedings of the committee were immediately transmitted to Governor General Gayoso, with a request to put a stop to Col. Grandprie's going into the district, if he should attempt it, as it would be considered a breach of the neutrality, and resented accordingly. This conduct alone, ought to be sufficient to convince every unprejudiced person, of the attachment of the permanent committee to the government and interests, of the United States.

Very soon after the part, which the permanent committee had taken relative to the admission of Col. Grandprie was publicly known, Mr. Hutchins entered into a correspondence with him, and it was currently reported in New Orleans that he had offered to aid him with two hundred men, in taking possession of the government. But the firmness of the committee, added to the general disposition of the inhabitants, rendered it improper, and perhaps unsafe for Col. Grandprie to make the attempt: in consequence of which he remained quietly at New Orleans, until he could be otherwise provided for. One of the letters from Col. Grandprie, to Mr. Hutchins, passed through my hands.

In the beginning of December, a considerable detachment of troops from the army of the United States arrived at Natchez: they had been long expected, and as long desired by the friends of the United States, as it was presumed their presence would have a happy effect, in keeping good order, by awing the turbulent, and rendering the possession of the country more certain. But it unfortunately happened, the Commandant [Captain Isaac Guion] who superceded Mr. Pope, was much indisposed by an inflammatory complaint on one side of his head, and face, at the time of his arrival, that evidently had an effect upon his understanding, which was naturally very far above mediocrity: in this state, he was immediately surrounded by a number of unworthy characters, who took advantage of his situation, to prejudice his mind against the permanent committee, and other friends of the United States; who were treated by him in the most opprobrious manner. One evening he came to the house which I occupied, and in which the committee held its meetings, and after insulting his Catholic Majesty's consul Mr. [Joseph] Vidal, who was there on a visit; he inquired whether or not the committee was in session? And being answered in the negative, asked if none of the members were present? Upon which Mr. Gaillard answered 'I am one of the members.' The Commandant then inquired 'by what authority they met?' That he was determined not to be made a cypher of, and would rule the district with a rod of iron. That the meetings of the committee were improper, and seditious, and must henceforth, consider itself dissolved, or he should treat it worse than Mr. Hutchins had done. Mr. Gaillard being a gentleman of considerable wealth, independence of sentiment, and firmness of mind, added to true republican principles, for which he had suffered under the Spanish government, could not brook the idea of again becoming subject to military despotism. He therefore replied with great warmth to the observations of the

Commandant, who after having exhausted his vocabulary of abusive epithets took leave with expressions too indelicate to have a place in language.[33]

Notwithstanding this outrageous conduct of the Commandant, the committee met the next morning without opposition; but the circumstance of his abuse of that respectable body, was taken advantage of by the discontented, who proposed that the district should immediately be put under military government. This, as might reasonably be expected, met with great opposition; and I considered it my duty, as a citizen of the United States, to prevent if possible, so dangerous and unnecessary a measure being carried into effect.

Among those who labored with the most assiduity in favour of the military government, was the late Mr. [Henry] Hunter, who was afterwards appointed by the influence of Mr. Hutchins, a member of congress from the Mississippi territory. Mr. Hunter was employed by me in May, 1797, to carry my despatches to the seat of our government, and returned a strenuous advocate for the omnipotence of presidential power, and pretended that he had received a commission from the United States, which enabled him to borrow a little money: the commission was probably a forgery. General [George] Mathews likewise exerted his talents on the same side of the questions with Mr. Hunter, but the behavior of the Commandant (already exemplified) on several occasions, had convinced every person in the district, who had the least regard to justice and liberty, that the measure would be attended with serious evils to individuals, and dishonor the United States. It therefore had an abortive issue.

In revenge for the failure of the above plan, my arrest was frequently spoken of, and as a foundation to ground it upon, an officer, with two other persons were despatched into the Chocktaw nation, to procure my letters; but as a cover to this expedition, a message or talk of Gen. Wilkinson, was

33. For a description of Captain Guion's antics and Ellicott's views of him, see Holmes, *Gayoso,* 232–33. "Ellicott and Guion are at drawn daggers, so that a House divided against itself can't last long on a solid and stable foundation" (Father Francis Lennan to Governor Gayoso, Natchez, January 31, 1978, trans. Jack D. L. Holmes, Archivo General de Indias, Seville, Papeles du Cuba, leg. 215-b. Ellicott's frustration with Lt. "Crazy" Pope, who was frequently intoxicated, Captain Guion's ego, and the known or rumored actions of General Wilkinson were summed up in a letter to Secretary Pickering, in which Ellicott wrote: "Is it possible to find a Military Gentleman in our army possessed of sobriety, talents and prudence? I have only to add that for the honor of the United States it will be necessary to send officers to this country who are not mad" (Commissioner Ellicott to Secretary of State, Natchez, July 4, 1797, in "Southern Boundary," vol. 1).

sent along. The day previous to the departure of those messengers, I met with the officer, and told him the expedition was unnecessary, as copies of all my letters were transmitted to the department of state, and if any of them should be found improper, or reprehensible, there could be no doubt but I should hear of it from the secretary of state, whom I could request to furnish them with copies, but if that would be attended with too much delay, they should willingly have the perusal of copies of them, in my possession. The officer was not accepted, and the party went, and returned, without procuring any materials for an arrest; but succeeded in persuading one of our deputy agents of Indians affairs by the name of [Samuel] Mitchell that I had prevailed upon the Indians to drive him out of the nation. A more bare-faced falsehood was never propagated!

For several months, the public mind was frequently agitated by anonymous papers that were circulated through the district; some of which were evidently written by Mr. Hutchins; but those that were entitled to the most attention, in point of composition, were penned by a Mr. Payne, and a Mr.*****; the latter for particular reasons, found it most convenient to go by an assumed name. Mr. Payne had followed trading and swindling on the Ohio, till his safety rendered it necessary to descend the river, and he arrived at Natchez a few weeks after I did: but it will only be doing justice to his character, to state that he voluntarily came in, and acknowledged his fault, and was afterwards useful in exposing to public view the insidious conduct of Mr. Hutchins. Mr. *****, had been a soldier in General Wayne's army, but after being twice publicly whipped for forgery, (at which he was very dexterous,) and other misbehaviour he was ignominiously discharged. He was a person of excellent understanding, and good education, but to all appearance totally lost to every sentiment of honour and honesty. Nature appeared to have intended him for an ornament to his country; but from a defect in the moral principle, he became the cause of sorrow to his connections. After being compelled to leave the army, he took his residence in the district of Natchez.

These two young men, almost strangers, without property, standing connections, or character in the country, became the champions of liberty, and the rights of the people. When men of this class become patronized for party, or political purposes, the public mind may be truly said to be taking in a slow poison, which like the forbidden fruit, does not produce immediate death, the effect is nevertheless certain, as well (to use the expression of the late

Gen. [Harry "Lighthorse"] Lee,) might the community be inoculated for the leprosy, as be governed or influenced by such characters. This sentiment I am well aware is not popular, neither was it in the republics of Greece and Rome. Those republics were constantly convulsed, and their political existence terminated by the violence of party, which was kept alive by artful demagogues, whose want of industry and character prevented their obtaining a livelihood by honest, and regular pursuits. When this class of men come forward as reformers, these questions ought to be asked. What interest have they in the community? 'Are they honest, are they capable'; what have they done to merit the confidence of the public? But it is said, that as liberty is as dear to one person as another, they are equally entitled to the use of their exertions, either in the obtaining or preservation of it. This is certainly true, so far as it goes merely to the preservation, or obtaining of liberty; but there may be an essential difference in their views: A person of character, and respectable standing in the community, and possessed of property has infinitely more at stake, than one possessed of neither; and consequently much more interest in keeping good order, and securing to individuals the unmolested enjoyment of their property. Many of those itinerant reformers, who have been falsely called politicians and philosophers, have under the specious garb of liberty, nothing but disorder and anarchy in view: they delight in confusion, for they are nourished and supported by it, like vermin by putrefaction.

Bad men, and fugitives from justice, think all governments however lenient, oppressive, they hate law, and are at all times ready to oppose it. It is therefore necessary to inquire dispassionately into the situation and probable views of every officious patriot, before we submit to become scaling ladders to his ambition, or instruments to impede justice, or prostrate government.

The very name of liberty, like that of religion, has in all ages entitled the true patriot, as well as the divine, to singular attention, respect and confidence; hence those characters, like valuable coin, hold out strong inducements to the artful and designing, to counterfeit them, from whence arises the necessity for honest men to be on their guard against them.

The permanent committee, by neither submitting to be guided or influenced in any of its measures, by the approbation of noisy or pretended friends, nor regarding the abuse of its opponents, gave such a stability to the public mind, that though the shadow of the Spanish jurisdiction which had remained in the district since the termination of the commotion, was withdrawn in January 1798, and the inhabitants left without law or government, till September following, I never heard of a single outrage being committed in the territory, except by the Commandant, and one or two other officers.

On the 18th of January 1798, the disagreeable state of suspense we had been in for almost one year, relative to the fate of the treaty, was partly terminated by the following letter from Governor Gayoso.

New Orleans, January 10th, 1798.

Sir,

By a packet just arrived, I have received orders from court, by which I am authorized and ordered, to evacuate the forts of Natchez and Nogales, (Walnut Hills,) in consequence thereof by this express, I send the necessary orders to withdraw the artillery, and other military effects. As the gallies will not be sufficient, other vessels shall be sent from this to complete the operation with all possible speed.

Please to furnish Major Minor with the information I request, that I may be enabled to provide everything concerning the execution of the boundary line between his Majesty's dominion, and the territory of the United States.

It is with the greatest satisfaction that I have the pleasure to announce to you this agreeable event, as it justifies our disposition in complying with our engagements as soon as political circumstances would justify it.

I have the honour to be,
With the highest consideration,
Sir, your most obedient humble servant,
Manuel Gayoso de Lemos.[34]

Honorable Andrew
Ellicott.

34. Manuel Gayoso de Lemos was appointed to the office of governor-general of Louisiana and West Florida to replace Baron de Carondelet on December 16, 1796. Gayoso announced his new appointment in July 1797, and in turn designated Captain Stephen Minor, who although American born had taken an oath of loyalty to the crown of Spain and served as an officer in the Spanish army with distinction, as the *ad interim* Spanish governor of Natchez. See Jack D. L. Holmes, "Stephen Minor, Natchez Pioneer," *Journal of Mississippi History* 42, no. 1 (February 1980): 17–26.

Immediately upon receiving the above letter, I had it made as public as possible, for the satisfaction our friends, and the same evening waited upon Capt. Minor, and gave him what information I could relative to arrangements and preparations necessary to be made on their part, agreeably to the request of Governor Gayoso, and the next day forwarded to him the following reply.

Natchez, January 19th, 1798.

Dear Sir,

Your favour of the 10th, came to hand yesterday in the evening. The intelligence which it contains is agreeable and interesting. Nations like individuals, can only be respectable in proportion to the good faith they observe in complying with their contracts: in this particular, the court of Madrid has generally been more exact than any other in Europe.

I have handed to Captain Minor, the arrangements made by the United States, for the demarcation of the boundary; but as this relates principally to the number of persons to be employed, we shall find it necessary on our meeting near the point of beginning, to settle many particulars which could not possibly be taken into view by the legislature and executive of the United States, owing to a want of the necessary information relative to the peace of the country, and many obstacles which could not reasonably be expected by persons only acquainted with the Atlantic states.

I am with great esteem and regard,
Your friend and humble servant,
Andrew Ellicott.

His Excellency Manuel
Gayoso de Lemos.

Although our business now began to assume a more favourable aspect, and had I not been taught by experience, to suspect everything that came from the officers of his Catholic Majesty, I should have concluded our difficulties at an end; but I still had my suspicions, which were in some measure realized on the 31st of January; when Capt. Guion handed me a letter which he had just received from Governor Gayoso, in which he informed the Captain, 'that he would come up to Natchez, and make the arrangements with him for furnishing the military escort, and supplying it with provision, likewise a plan for running the boundary.' Upon reading this curious letter, it was observed to the Captain, 'that delay and embarrassment, still continued to be an object with Governor Gayoso; but the present finesse would, and must fail.' The Captain replied, that he 'was certainly the proper person to make the arrangements, but did not intend to interfere with the astronomical operations.' Finding from this reply, that the Governor's finesse had taken with the Captain, the following letter was sent off to New Orleans, for the Governor the next day.

Natchez, February 1st, 1798.

Dear Sir,

That our proceedings in carrying into effect, that part of the late treaty between the United States and his Catholic Majesty, which has been submitted to us, may be commenced with system, it will be necessary that we should understand each other as to the place, and precise time of our meeting; by the third article of the treaty, my present situation appears to be the place contemplated by the contracting parties for the meeting of the commissioners; however as that cannot be very material, I shall have no objection to meet you at Clarksville, which will be near the beginning of the line of demarcation. With respect to the time I wish it to be at as early a period as possible; and as I am at times ready to proceed to business, I wish you to name the day of meeting.

Captain Guion has communicated to me the content of a letter, which he has lately received from you relative to the arrangements for furnishing the military escort, supplying the provisions, and a plan for carrying on the line. With respect to the first of those objects, I beg leave to observe, that the escort has already been furnished by Colonel Butler, agreeable to instructions which he received from the secretary of war [James McHenry, 1796–1800]: of this circumstance you could not possibly be unknowing, as the escort descended the river with me, and was for some time a subject of discussion between us. As to the second point, the supplying of provisions, it is equally well known to your Excellency, that a commissary has been appointed by the secretary of state of the United States, for the purpose of procuring the necessary supplies, and attending to the transportation of provisions, instruments, etc. So far, I thought the arrangements would be considered without the reach of criticisms or objections, and I confess that I am unable to account for your letter to Captain Guion upon any other principle, than the contemplation of further delay. The plan, which you have proposed to the same gentleman, for carrying on the line, however proper and judicious in itself, is wholly unnecessary, as he is not the person appointed to carry that part of the treaty into effect. My instructions, and those to the surveyor, I am fully persuaded will be a sufficient guide to us in the execution of the business, without any foreign or domestic advice.

I now give you my candid opinion, that I think the arrangements on behalf of the United States, fully sufficient to give validity to the work when executed, and if yours should not be much more exceptionable, there will be no objection from the government I have the honour to serve, which would spurn at the idea of criticism on extraneous minutia, which could neither affect the accuracy, nor due execution of the work.

I have the honour to be,
With great esteem and respect,
Your sincere friend and humble servant,
Andrew Ellicott.

His Excellency Manuel
Gayoso de Lemos.

To the foregoing letter Governor Gayoso returned the following answer.

New Orleans, February 13th, 1798.

Sir,

I have received your favour of the 1st instant, by which you propose our meeting at Clarksville, as soon as possible, finding yourself in readiness to proceed to the demarcation of the boundary line prescribed by the treaty.

In consequence of the measures that I have taken for evacuating the posts, I am persuaded that some time next month that the operation will be completed, at the same time I shall be at Clarksville, disposed to begin the demarcation of the boundary line; but previous to my arrival, I shall give you timely notice, that we may meet about the same time.

My communication to Captain Guion on the same subject, ought not to appear strange to you. General Wilkinson speaking of Captain Guion to me says, 'this officer's experience and good sense, the powers with which he is elevated by the President of the United States, conspire to promise a happy result to his command, in which I flatter myself, I shall not be disappointed.' Under such circumstances Captain Guion is not an indifferent person to me, and the double application I have made to you and to him, does not imply the disposition that you are pleased to interpret; the executive of the United States can give his powers either to you, or to Captain Guion, or to both, and where I find that the authority resides, I have no manner of objection to act with it, for in this important business, the depending interests are those of the United States, and the king my master executed by their representatives.

I am, with great esteem and
Friendship, sir,
Your humble servant,
Manuel Gayoso de Lemos.

Honourable Andrew
Ellicott.

The latter part of the above letter from Governor Gayoso is certainly very confused, he must have been sensible, that his finessing with Captain Guion was highly improper, and could not be justified by any character given of him by Gen. Wilkinson: he therefore found it difficult to reconcile his conduct with that candour he had constantly professed.

Not being by any means certain of the extent of Mr. Blount's plans and intrigues, (a friend of his being then in the settlement, and in the pay of the United States,) or in what manner they might operate in embarrassing our business, or how far those delays were in unison with another plan, of much greater magnitude, the outlines of which I obtained in October 1797, and communicated confidentially in cypher to the department of state, by a communication dated the 14th of November following, it was therefore resolved, to bring the business to issue at all events, and risk the consequences, by beginning the demarcation of the boundary, the instant the posts were evacuated by the Spaniards; although the officers of his Catholic Majesty should not co-operate, and Captain Guion should detain my escort. This resolution was not made known to an individual in that country; but communicated to the department of state by two despatches, the one dated February [14th] 1798, and the other the 2th of the same month.

The French privateers had now begun to be very troublesome to the trade of the United States in the West Indies, and about the Gulf of Mexico. A number of our captured vessels were taken into the port of New Orleans, condemned, and disposed of with their cargoes for a trifling price, our seamen treated in a most shameful manner, and our trade otherwise brought into great jeopardy.

This subject became a matter of serious consideration, and the United States having neither consul nor vice consul at that port, I interested myself in procuring the privilege for Daniel Clark, junior [a nephew of Daniel Clark Sr., planter of Natchez], a distinguished merchant of that place, to act as consul for the United States, till the executive should make a regular appointment. Immediately upon the application being made to Governor Gayoso, he directed that Mr. Clark should be received as consul from the United States, and regarded as such by the merchants and officers of his Catholic Majesty.[35]

The firm, manly, and decided conduct of Mr. Clark, in a short time put a new face upon our commerce in that quarter, and obtained some privileges we had not before enjoyed. The correspondence between Governor Gayoso, Mr. Clark and myself, on the various subjects which invited discussion, would of itself make a volume; but as it more particularly concerned Mr. Clark, it is hoped he will some time make it public. The conduct of Mr. Clark was so acceptable to the executive of the United States, that he was thanked through my hands, and requested to continue his good offices in favour of our citizens; the appointment of consul was nevertheless given to a Mr. [Evan] Jones, and that of vice consul, to a worthy native of the United States, by the name of

35. On the role of Daniel Clark Jr., vice consul for the United States in New Orleans, see Clark, *New Orleans 1718–1812*, 215, 240, 242–45. In a dispatch to Secretary of State Pickering (November 8, 1798), Ellicott added a postscript that stated: "Daniel Clark, Esq. of New Orleans, has lately spent a number of days with me in camp. From him I have received much valuable information, which it will be unnecessary for me to detail, as he will give it to you himself in Philadelphia, the ensuing winter. He intends to visit that city, immediately after an interview in New Orleans" (qtd. in Wilkinson, *Memoirs*, vol. 2, appendix 31).

[William E.] Hulings. Upon the late change of administration taking place, the merit of Mr. Clark was so conspicuous, that he was appointed consul in the place of Mr. Jones.[36]

During the years 1797 and 1798, the character of the United States stood very high in the Floridas and Louisiana, and even those who were directly in favour of the republic of France, admired the firm manner in which Mr. [John] Adams began his administration; and could not withhold their approbation of the threatening aspect assumed by one of the youngest nations in the world, that dared boldly to come forward, and oppose a power under which all Europe was bending. This conduct produced such a confidence in the friends of the United States in that quarter, that a plan was early formed, to add to the Union, the two Floridas, with the island of Orleans, provided the Spaniards either committed hostilities against the citizens of the United States at Natchez, or joined France in the contest against us.[37] From the

36. Clark, *New Orleans 1718–1812,* 214–15, 245–46.

37. Cox describes the situation in Natchez at this time: "President Adams's determined attitude toward France so aroused the American contingent at Natchez that they formed a plan to add to the Union 'the two Floridas with the Island of New Orleans,' in case the Spaniards began hostilities, or permitted the French to move through their territory" (Cox, *The West Florida Controversy, 1798–1818,* 45). Additional details on the tense situation are provided by James Pitot (1761–1831), who was in Natchez. Pitot was a French immigrant to the United States, where he received American citizenship in 1796 as a Philadelphia merchant. Upon the opening of Spanish province of Louisiana to Americans, he decided to settle there; in the process his travels took him through Natchez during the years of turmoil, 1797–98. On American interests, he wrote: "Under the pretext of uprisings among the Indians and the landing of munitions and soldiers by the French on the Florida coast to help them, the United States recruited the little army in their eastern states which they rapidly moved into the Mississippi Territory [established in 1798], and with which they destined to make Spain regret her culpable policies toward the province of Louisiana. In this manner they acted on expectations, all the more justified as perhaps I shall not exaggerate too much in saying that the colony found itself half surrendered because of its situation to the person [General James Wilkinson arrived in Natchez in August 1798] whom Spain bribed several years before to condition favorably the sentiments of the Kentucky inhabitants, and governed by the man [William Blount] who, when he held a lesser rank [as governor of the Southwest Territory (1790–96)], disseminated the irresistible oratory and arguments of desertion. By reciting these events, virtually proven in the eyes of all observers, I do not flatter myself that I have unimpeachable proofs in my possession; but in addition to agents and eyewitnesses to many of these intrigues who have given me details confidentially, it is a fact that I was in the Natchez country in 1798, when the recruits and general of the little army arrived" (Pitot, *Observations,* 16–17). Pitot's *Observations* were written in 1802 with the intention of providing them as a report to the French

secrecy, talents and enterprise of those concerned, added to a temporary system of finance, and deposit of arms, there could not possibly be any doubt of the complete, and almost instantaneous success of the plan had it been attempted. But the threatening aspect which was assumed by our government, like a meteor, flashed but for the moment, left the nation paralyzed and damped, or extinguished the spirit of enterprise, without obtaining a single permanent advantage.

If Mr. Adams was right in the manner he commenced his administration, he was wrong in abandoning the ground he had taken: but if he began wrong, his subsequent conduct was by no means calculated to do credit to the country, or give the shadow of consistency to the measures of government, which were so capricious, that they left neither the French nor British satisfied with us, who may hereafter combine their intrigues and forces, to prostrate the only free national government on earth.

The existence of the plan above mentioned, for annexing the Floridas and the island of Orleans to the United States, is mentioned in a long communication to the department of state, dated January 10th, 1799. A further public development would for very obvious reasons, be yet improper.

Being fully satisfied, that delay was yet a favourite object with the officers of his Catholic Majesty, and having taken the resolution already mentioned, to commence the business without them so soon as the posts were evacuated; I waited in daily expectation of that event. The time when the evacuation was to take place, was kept a most profound secret by the Spaniards, and managed with all the caution of a retreat. On the 29th of March late in the evening, I was informed through a confidential channel, that the evacuation would take place the next morning, before dawn; in consequence of which, I rose the next morning at four o'clock, and walked to the fort, and found the last party, or rear guard just leaving it, and as the gate was left open, I went in, and enjoyed from the parapet, the pleasing prospect of the gallies and boats leaving the shore, and getting under way: they were out of sight of the town before daylight. The same day [30 March 1798] our troops took possession of the works.

---

government as it prepared to accept the province of Louisiana from Spain, and with which he hoped to establish a good relationship. According to Whitaker, Ellicott was directly involved in the effort of encouraging settlers to abandon Spain (Whitaker, *The Mississippi Question,* 61).

CHAPTER VI.

The author leaves the town of Natchez—difficulties which attended the beginning of the boundary—joined by Governor Gayoso and other officers of his Catholic Majesty—Governor Gayoso returns to New Orleans and sends back a communication concerning the hostile disposition of the Indians—Opinion of it—moves on by several stations to the (Pearl) or Half Way River—on the way gets possession of a curious letter—difficulties at the Pearl river—finishes the course of observations, and sets out for New Orleans—arrives there on the 4th of January 1799—some account of the Pearl, or Half way river, and lake Pontchartrain.

I now began to make my arrangements for commencing the important business of my mission, and left the town of Natchez the 9th of April following.

I confess my feelings were much alive on leaving that town: the attention, politeness, and hospitality which I had experienced from the inhabitants on all occasions, for more than a year, had made strong impressions upon my mind, which can only be obliterated by the loss of recollection. And though it was frequently the scene of difficulties, the constant support I found from the friends of the United States, and ultimate success of the important business with which I was intrusted, served to render the place more interesting. Our boats were under way by sunrise; but I could not direct my face from the town, until the last house disappeared.

On the 10th early in the morning, we arrived at Clarksville and encamped, and on the 11th I set up the clock, and small zenith sector, and proceeded to take the zenith distance of pollux [a star in the constellation Gemini], for five evenings successively, the first three, with the plane of the sector to the east, and the others with the plane west. From the result of those observations, it appeared that we were three miles, and two hundred and ninety perches [a unit of length equal to 16.5 feet] too far north. This distance my assistants Messrs. Gillespie, Ellicott, junior, and Walker, traversed with a common surveying compass and chain, to the south, in order to discover, (nearly,) a proper place to encamp, and set up the large sector, to determine the first point in the line with accuracy. When this traverse was completed, it was found to be impracticable to convey our instruments, baggage and stores directly from Clarksville; to the most eligible place, owing to the extreme unevenness of the country on the one hand, and the banks of the Mississippi not being sufficiently inundated on the other, to give us a passage by water through the swamps and small lakes: it was therefore determined to descend the Mississippi to the Bayou Tunica, (or Willing's Bayou;) from whence I understood we could convey our instruments, stores and baggage, either by land or water; almost to the place of beginning; though not without some difficulty. The distance from Clarksville to the Bayou Tunica by land, is about eight and an half miles, but by the Mississippi more than fifty.

On the 24th we left Clarksville, and arrived at the Bayou Tunica on the 26th, being detained one day by head winds.

On the 27th my assistants were sent to carry a line east from the termination of the travers already made, into the high land; and on the 28th, I went and examined the country over which the guide line passed, and fixed upon a very elevated situation; about one thousand four hundred feet south of it, for our first position: but the difficulty of getting our instruments, baggage and stores to it, appeared much greater than I first expected. A party of our men were directed to open a road from the height already pitched upon, to Alston's Lake; this distance was about one mile. The road was completed on the 30th, and on the first day of May we moved and encamped on the top of the hill. Our instruments, baggage, etc., were first carted from Bayou Tunica, to Alston's Lake, into which I had previously taken through the swamp two light skiffs: the articles were then taken by water, up the lake to the point where our road from the hill struck it, and from thence packed on horses to our encampment. The county was so broken, and covered from the tops to the bottoms of the hills, with such high, strong cane, (arundo gigantea,) and a variety of lofty timber, that a road from the Bayou Tunica, to our camp, could not be made by our number of hands, in less than a month passable for pack-horses.

The situation of our encampment on the top of the hill, was both pleasant and beautiful, the prospect fine, particularly to the south-west, which opened to the Mississippi swamp, and gave us an uninterrupted view, which was only terminated by the curvature of the earth. The hill appeared to have been under cultivation some years ago, and the cane entirely destroyed; but a number of splendid trees were left standing, among which we encamped.

Our observatory tent being worn-out by the military, who had no tents when they arrived at Natchez, I was now under the necessity of erecting a wooden building for that purpose; which I began on the 2nd of May, and with the aid of four men finished on the 4th, and set up the clock, and large zenith sector; but the weather being unfavourable, the course of observations was not began till the 6th, and was completed on the 16th , which with the results, and manner of commencing demarcation of the boundary, may be seen in the Appendix. On the 21st we were joined by Capt. Minor, and a party of labourers on behalf of his Catholic Majesty, and on the 26th by Mr. [William] Dunbar, the astronomer in the same employ.[38]

On the 31st of May, governor Gayoso, with his secretary and several other officers arrived at our camp. The Governor had written me a number of letters to delay any proceedings till he joined me; but being well convinced, if it had been left to that issue, he would have found reasons for continuing at New Orleans the whole season.

On the 1st of June, the Governor accompanied me to the line, and approved of the work. On our return to camp, I requested him to join me in a report confirming the work already done, particularly as he was yet the principal commissioner, and had not under the powers with which he was invested by the court of Madrid, authorized any other person to execute the commission; but under various pretences, he declined doing. And the day following set out for New Orleans, after impowering Capt. Minor and Mr. Dunbar, to execute the commission on behalf of his Catholic Majesty.

On the 7th of June, we moved our camp to Little Bayou Sara. The weather had now become extremely hot, and the season being uncommonly wet, and our men badly provided, for tents and other covering, they were generally indisposed and unfit for duty. We were therefore reduced to the necessity

38. See DeRosier, *William Dunbar*; Jack D. L. Holmes, "William Dunbar's Correspondence on the Southern Boundary of Mississippi, 1798," *Journal of Mississippi History* 27, no. 2 (May 1965): 187–90; and Gene Allen Smith and Sylvia L. Hilton, eds., *Nexus of Empire: Negotiating Loyalty and Identity in the Revolutionary Borderlands, 1760s–1820s* (Gainesville: University Press of Florida, 2011).

of either abandoning the business for some time, or employing slaves; the latter was adopted.

The first twenty miles of country over which the line passed, is perhaps as fertile as any in the United States; and at the same time the most impenetrable, and could only be explored using the cane knife and hatchet. The whole face of the country being covered with strong canes, which stood almost as close together as hemp stalks, and generally from twenty to thirty-five feet high, and matted together by various species of vines, that connected them with the boughs of the lofty timber, which was very abundant. The hills are numerous; short and steep: from those untoward circumstances, we were scarcely ever able to open one-fourth of a mile per day, and frequently much less.

On the 10th of June, an official communication was received from Governor Gayoso; informing us of the hostile disposition of the Indians, and their determination to put a stop to the demarcation of the boundary; and that their enmity was directed against the people of the United States: this intelligence was considered as part of the system of delay, which had already been so injurious and expensive to the United States, it was therefore disregarded; but from an impression of duty, the communication, with some observations upon it, was forwarded immediately to the department of state.

On the 17th of July, we moved our camp to Big Bayou Sara; but a few days previous to our removal, we received information that a constitution had been framed for the Mississippi territory, and that Winthrop Sargent, Esq. was appointed Governor.[39] This intelligence was not only agreeable to us, but highly acceptable to every friend to order and good government in the territory. We were further informed, that General Wilkinson was expected down to take command of the troops.

Before the line was extended to the Bayou Sara, Mr. Hutchins, and several others made application for lands south of the line, and requested permission to remove to them, but Governor Gayoso denied the request to Mr. Hutchins, and some other turbulent characters; but permitted one of Mr. Hutchins's sons to remove within his government, where I am informed he continues to reside: those turbulent characters finding now no alternative left, they then,

39. Winthrop Sargent (1753–1820) was a Federalist from Massachusetts who had previously served as secretary of the Northwest Territory. On the situation in Mississippi, see the excellent overview provided by Haynes, *The Mississippi Territory*.

and not till then, become violent republicans, and may now be said to direct generally the public affairs of that territory.

On the 6th of August Governor Sargent arrived at Natchez, but in so bad a state of health, that he did not begin to organize the government till in September following. On the 26th of the same month General Wilkinson arrived.

Thus it will appear, that all the difficulties we had to encounter were surmounted, and the demarcation of the boundary commenced several months before the arrival of either the Governor or General Wilkinson.

In the beginning of September Mr. Dunbar left us, and returned to his seat, a few miles from the town of Natchez. This circumstance I considered a real loss to the public. To myself it was irreparable.[40] The day after the departure of Mr. Dunbar, we moved to Thompson's Creek, where we continued till the 27th of October, when we broke up that camp, and proceeded to the Pearl, or Half-Way River: on the way I halted a few days at Darling's Creek, for the purpose of refreshing our pack horses, and making up a communication to the department of state. While at that place, by a very extraordinary accident, a letter from the Governor General [Gayoso], on its way to a confidential officer in the Spanish service, fell into my hands for a few hours.

This letter contained the most unequivocal proof, of the late existence of a plan, calculated to injure the United States; but which appeared then to be abandoned, and in which a number of our citizens had been actually engaged. From this letter it was rendered evidence, that the suspicions of the late General Wayne, respecting an improper correspondence being carried on between the officers of his Catholic Majesty, and some gentlemen residing in the western part of the United States, were well founded; but it was equally certain, that he was mistaken in the several cases, as to the individual dual

40. William Dunbar, as astronomical commissioner for Spain, took ill in July, and this, combined with an issue over his possible conflict of interest in serving both Spain and the United States, led him to request that a replacement be appointed. Ellicott wrote Secretary Pickering (July 29, 1798): "As soon as Mr. Dunbar completes this part of the business [the survey line] we shall confirm the work: Mr. Dunbar will then retire from the service, and leave Major Minor the sole commissioner on the part of his Catholic Majesty. Mr. Dunbar's retiring will leave a heavy burden on my shoulders, his scientific talents would be useful in any country, and command respect in any government. His retiring is owing to a delicacy he feels as a citizen of the U.S. and at the same time acting for, and receiving pay from another power." Governor Gayoso named Major Stephen Minor to replace Dunbar and to complete the boundary line with Ellicott.

objects of his suspicion. It is likewise a fact, that the despatches, nearly twenty thousand dollars in silver, to be paid to certain characters were on board of the boat from New Orleans, which was taken by his order, and examined by Lieut. Steel; but the articles were overlooked.

The interesting parts of the above mentioned letter, were reduced to cypher, and accompanied my despatches of November 8th 1798, to the department of state.[41]

On Saturday the 17th of November, we arrived at the Cane Brake on the west side of Pearl river, and the 18th were employed in opening a road to the river, making rafts, and ferrying our baggage over, and on the 19th encamped and set up the clock.

In our journey from Thompson's Creek, we had many difficulties to encounter. The swamps were numerous, and many of them so deep, that we had to go considerably out of way either to cross; or go round them, and others we had to causeway: add to those difficulties, a total want of information respecting the face of the country, which in our direction, did not appear to have ever been explored by white people; some of the streams were so deep, that we had to cross on rafts, in the constructions of which my assistants already mentioned, were singularly useful.

From Thompson's Creek east along the boundary, the soil is of an inferior quality, except on the margins of creeks and rivulets, which is very fertile, and covered either with large cane, (arundo gigantean,) or the small cane or reed,

41. Ellicott wrote a lengthy report on the subject (November 8, 1798) to Secretary of State Pickering: "Mr. [Thomas] Power, a gentleman well known for his intrigues in Kentucky, and other parts of the United States, is the surveyor on the part of the crown of Spain: he has attended but one week on the line; and I do not believe that he will attend another during the execution of the work. . . . There are no Spaniards concerned in the business, but a few common soldiers. Major Minor and Mr. [Daniel] Burnet, are Americans; the others, including the labourers, are generally French, or descended from French ancestors, or Roman Catholic Irish. When I look over this strange heterogeneous collection, I cannot help asking this question, 'Can the Spaniards really be serious in carrying the treaty into effect?' If they are, it is very extraordinary, that there is not one of that nation employed, above the rank of a common soldier." Ellicott then reiterated his earlier communications regarding the delaying tactics of both Baron de Carondelet and Gayoso and concluded with his views on the French: "For my part, I have expressed so much want of principle and integrity, among them and their partisans in this country, both individually and collectively; that my prejudices against the whole nation is so strong, that it is with difficulty I can guard my expressions, so as not to give offence" (Andrew Ellicott, "Extracts from A. Ellicott's Communications to Secretary Pickering," qtd. in Wilkinson, *Memoirs*, vol. 2, appendix 31).

(arundo tecta.) The fact of the country is gently waving with hills, which in some places have a most beautiful appearance. The prevailing timber on the upland is pine, (pinus,) of several species. The grass is high, coarse and hard, and when full grown is only eaten by horses and cattle, through necessity. The soil of the upland is composed of a large portion of sand, with a small mixture of yellowish clay or loam. On one of the pine ridges I saw a few stones, which were evidently ferruginous; but this appearance was not promising.

The day on which we encamped, a number of our men were set about making a canoe, to descend the river, and meet our provisions that were expected from New Orleans; the small supply we had brought on from Thompson's Creek being near exhausted.

On the 21st the canoe being finished, Mr. Robins the superintendent of our labourers, with three hands were despatched down the river; but returned on the 23rd, in company with a small, light skiff, which had been sent up from the mouth of the river to examine the state of the navigation. The hands who had the care of the skiff, informed us that our stores and the large sector, were at the Bluffs above the mouth of the river, but could not possibly be brought up, till a large quantity of timber was removed, and channels cut through two rafts or floating bridges, which extended from one side to the other; and those rafts or bridges, being in the swamp, the articles could not be carried round them.

On the 24th, my assistant Mr. Gillespie, with a number of our labourers were despatched down the river to remove the obstructions; from there he was to proceed to New Orleans, and make arrangements for an immediate supply of some articles we were in want of. Our provisions, (beef excepted,) were exhausted on the 27th.

Having two more canoes finished on the 28th, the superintendents of both parties, were sent down the river with some of our most active men, to bring us a supply of flour with all possible expedition.

On the 30th a small supply of provisions arrived on pack-horses, from Thompson's Creek. The pack-horse men likewise brought on my small zenith sector. This instrument was not intended to be made use of in the determination of any points in the boundary, and merely brought along for common geographical purposes; but owing to the uncertainty of the time when the large one would arrive, which was too weighty to be conveyed by land, and therefore sent round by water, added to the fickleness of the weather at that season, and being in want of some of the necessaries of life, it was thought proper and expedient to use the small one. The 1st of December, I polished the mirrors of the eye piece, which were tarnished by the great moisture of that country, and set the instrument up.

The weather was cloudy until the 3rd, when a course of observations was begun, which was closed on the 13th, much more to my satisfaction, than could reasonably be expected from so small an instrument. From the time these observations commenced, until they were completed, I had but few hours sleep; for the space of three successive days and two nights, I slept but three hours.

The day on which the observations were closed, Mr. Gillespie arrived from New Orleans, and brought with him a few barrels of flour. His expedition on this occasion, and the manner in which he executed his trust, do him the greatest credit.

On the 15th and 16th, we were employed in laying of the correction, and making the necessary arrangements for carrying on a guide line to the Mobile, and correcting back to Thompson's Creek. The former was submitted to Mr. Daniel Burnet, and the latter to Mr. Gillespie.

The difficulty of passing through the country from Thompson's Creek to the Pearl river, was so great, owing to the swamps and morasses which lay in the way, and were daily filling with water, that the principal part of our baggage and clothes, which we had left behind, had not yet overtaken us. I nevertheless found it necessary to set out immediately for New Orleans, with no other clothing than what was immediately calculated for the woods. This measure had become necessary for the following reasons. It has already been mentioned that our tents were worn out before we commenced our business, and myself and people without the requisite covering. The escort was without either pay or clothing. On these subjects I had early written to the department of state, and received assurances that the articles should be immediately forwarded; but from some cause or other, it was never done. We had therefore to procure them at an advanced price in New Orleans. It has likewise been observed, that Governor Gayoso left us without officially confirming the work that was executed before he joined us, his signature on the report was yet wanting, and it did not appear probable that it would be had without waiting upon him. Experience had already taught us, that it would be impossible to convey our apparatus, baggage and provision wholly by land; a vessel to carry the heavy articles was therefore wanting, calculated to follow the coast, and ascend the navigable rivers, to or near points where the line demarcation crossed them. With this, another object which appeared of considerable importance was combined, which was to obtain as accurate a knowledge as possible of the sea coast, and of the navigation of the rivers that rise in the United States, and fall into the Gulf of Mexico.

On the 17th in the afternoon, I left my encampment, accompanied by a part of the military escort. Our passage down the Pearl river was extremely disagreeable, the rain was cold and almost incessant, and our canoes small and uncovered, which added to other obstacles, prevented our reaching the Bluffs, on the tide water, till the 25th. From the Bluffs a person was immediately despatched to New-Orleans, to procure a vessel to convey myself and party across Lake Pontchartrain, which by the exertions of Governor Gayoso, was immediately obtained, and arrived at the Bluffs in the forenoon of the 1st of January 1799. On the same day about three o'clock in the afternoon we got under way, but the tide being against us, we came to an anchor about six o'clock in the evening. Early the next morning we were under way, but the tide failing, and the wind being ahead, we came to an anchor in the Rigolets, where we lay until the morning of the 3rd; when we got under way, and entered Lake Pontchartrain about ten o'clock in the forenoon, and having a light, fair breeze, we arrived at the fort, at the mouth of Bayou St. John's in the evening. Early the next morning, being the 4th, the Governor's barge was sent to carry me to New Orleans (the distance was about six miles,) where I arrived at ten o'clock in the forenoon, and was received by the Governor and officers of the government, in the most polite and friendly manner. I shall now close this chapter with some account of the Pearl river.

The Pearl or Half-Way River, is navigable for small craft many miles north of the boundary. It is remarkably crooked, and full of logs and lodged trees; which are at present very injurious to its navigation. Its banks for some distance above the boundary, and almost the whole of them below, are annually inundated. The banks, with a considerable extent of country become very low, below the Indian house, (marked on the map,) over the whole of which the water passes, when the river is high; and here it begins to divide into a number of branches, some them maintain an open channel until they unite again with the main branch; and others are lost in the swamp. Those branches appear so nearly of the same size, that a person not acquainted with the river, will be as likely to take a wrong, as a right one. This happened to several of our parties and to myself, although I had two persons with me, who had been up and down twice before: we were a part of two days and one night, before we got back to the place where we made the mistake. The officer of my escort, with several of his men, were still more unfortunate; they took another branch, and were a greater length of time before they discovered their error, and on half allowance of provision.

In consequence of the water extending over such a considerable space, it never acquires a sufficient head to force away the lodged timber, which in two places extend across the river.

The upper raft [logjam] is of considerable magnitude, and covered with grass and other herbage, with some bushes. Through those rafts our man had to make channels, by removing and cutting away the logs, until they had a sufficient depth of water to float loaded periaguas [pirogues] and canoes. This was an arduous undertaking, and executed at the most unfavourable season of the year.

The tide ebbs and flows, a few miles above latitude 30°21′30″, where there was formerly a trading house, and where any vessel that can cross the bar into the Lake, may ascend with ease. The banks of the river above the old trading house, so far as the tide is perceptible, are too low and marshy for a settlement. The river has several communications with the Gulf of Mexico and Lake Pontchartrain, but they are all too shoal for vessels drawing more than six or seven feet water, and therefore only fit for the coasting trade.

The coasting vessels which visit New Orleans from the eastward, pass by the mouth of the Pearl river into Lake Pontchartrain, thence through the west end of the Lake, and up the Bayou St. John's to the canal, executed by the Baron de Carondelet, thence to the end of the canal, which terminates at the walls of the city, immediately behind the hospital [a route which later became Canal Street]. This canal requires cleaning every year, which is done by slaves and criminals, condemned to hard labour; but might be done more effectually, by conveying a stream of water into it from the Mississippi, at the time of the annual inundation, which might be effected with but little trouble and expense.

Lake Pontchartrain which connects the mouths of Pearl river, with Bayou St. John's and thence with New Orleans, is a beautiful sheet of water, but unfortunately surrounded with marshes, and the landing in many places is attended with difficulty on account of the mud. There are some places towards the east end, where the beach has a fine appearance, being composed of large bodies of cockle shells, from which the lime is used at New Orleans, and about the Lake is made. The water in the Lake is not deep, being generally not more than twelve or fifteen feet.

CHAPTER VII.

The author makes a course of astronomical observations—obtains a vessel—some account of the city of New Orleans—ceremony at signing of the reports—proof of Mr. Hutchins being at that Time a British officer, with some remarks on that subject—leaves New Orleans, and arrives at the guide line on the Mobile river—an observation relative to the Pascagola river.

Being anxious to examine the geographical position of the city of New Orleans, the large sector, telescope and clock were unpacked, and set up on the 10th and 15th of January, and a number of observations made during my stay, which with the results will be found in the Appendix.

Immediately on my arrival in New Orleans, arrangements were made for procuring the articles we were in want of; but it was found difficult to obtain a vessel properly constructed, to answer our purpose on the coast, and for ascending the rivers. A hull, without a deck was a length procured, but of live oak and red cedar; it was very light; but well put together. On the 24th of January several hands began to work upon her, whom I superintended myself from daylight until dark, until she was ready for sea, (Sundays not excepted,) for which permission was had from the bishop.

New Orleans has now become a place of very considerable importance, both on account of its population and commerce, and some gentlemen of respectable talents are looking forward with pleasure to a period, which they conceive no distant, when it will be annexed to the United States. For my part, I do not see any advantage we could derive from the possession of it at present. The United States are already in a great degree possessed of its commerce, and draw from it annually a very large sum in specie, and that probably, with much more ease than if it was in our possession. When I give this opinion, I would be understood to mean while it is in the possession of his Catholic Majesty. Rather than a transfer should be made of it to any power in Europe, or than it should become a part of a new empire, I should think it our interest to possess it.[42]

It has been doubted whether the local situation, or site of the city of New Orleans, is the best that could be chosen to combine generally the greatest number of advantages, with the fewest disadvantages. Lower down the river would be more convenient as a sea port; but the ground is lower, softer, and the country less healthy: further up the river, the country is higher, firmer, and more healthy; but the difficulty of ascending with shipping, would increase almost in geometrical ratio. For it must be observed, that although there is a small swell in the river, which is sometimes perceived as high as New Orleans, the current is nevertheless always strong into the Gulf of Mexico; and therefore, no advantages are ever to be expected from the tides. And when the winds have been unfavourable, vessels have been known to be upwards of six weeks in going up to New Orleans from the Balize [the pilot station at the mouth]; which is a serious drawback upon the profits of a voyage; add to this the danger of sickness among the hands, if they should be unfortunately delayed in that low, marshy country in summer, or the beginning of autumn, and it will probably appear, that the city is already sufficiently distant [more than 60 miles] from the mouth of the river. There is one argument in favour of the present situation which has not been answered by the advocates for a position higher up, and that is, the facility with which all the coasting trade east of the Mississippi, is connected with the city of New Orleans, by means of the canal already mentioned, the Bayou St. John's and the Lakes, and which could not be carried on with the same ease at any other point. However if the situation is not the best, it is now too late to remedy it as the wealth, population and capital in trade is so considerable, that the certainty

42. Ellicott believed at this point that American commerce, using the port of New Orleans even under the Spanish, resulted in lucrative commerce for the United States, and only if it were to pass to another European power, by whom he meant France, would it then be desirable to take it by force into the Union. It was a common assumption at this time that a "feeble" Spain was under the control of France and that, as Senator William Blount had feared, the province of Louisiana would be ultimately retroceded from Spain to France. This is precisely what happened on October 1, 1800. On this topic, see Mildred S. Fletcher, "Louisiana as a Factor in French Diplomacy from 1763 to 1803," *Mississippi Valley Historical Review* 17 (December 1930): 367–76; and Arthur P. Whitaker, "The Retrocession of Louisiana in Spanish Policy," *American Historical Review* 39 (April 1934): 454–76.

of a market will prevent any competition for many years, and consequently impede the growth of any other place within a reasonable distance.

No place upon this continent, and perhaps in the world, can command the trade of an equal extent of fertile country as that of New Orleans; and as that vast country increases in population, so must that city in magnitude, wealth and commerce.

The weather during the summer season at New Orleans is warm, sultry, and disagreeable, but during the cool months, there are few places more desirable: it then abounds with health, and a variety of well conducted amusements, which are encouraged and protected by the government; but this, though pleasing, it may be observed, is characteristic of despotism, and naturally grows out of an arbitrary government. The mind of man being active, must be employed, and if not occupied by amusements, may in its pursuit of objects to rest upon, be directed to the investigation of the principles of liberty, and inquiries into the conduct of public officers, which are of all things the most to be dreaded, and are the most exceptionable to the feelings of an arbitrary magistrate.

The plan of the city of New Orleans is regular; the streets cross each other at right angles; and are accommodated by their narrowness, to the heat of the climate. It is fortified on the sides exposed to the land by a work, which though not strong, is far from being contemptible.[43]

The city has suffered several times severely by fires, but has entirely recovered from the effects of them; by the last [1794], the greater part of it was laid in ashes. It now affords many good houses, built in a handsome style.

On the 23rd of February, Governor Gayoso and myself signed four reports, two in English, and two in the Spanish language, confirming all the work done before the 7th of June, 1798; after that day, the execution of the work on behalf of his Catholic Majesty, was submitted to William Dunbar, Esq. and Capt. Minor; but Mr. Dunbar continued by a few months in the employ, after which the whole duty devolved upon Capt. Minor. One report in each language was intended for the executive of the United States, and the other two in the same manner for his Catholic Majesty.

Great ceremony was used at the signing the instruments. The Governor

43. The plan of New Orleans provided fortifications on both the riverfront and landward corners—Fort St. Charles, Fort St. Philip, Fort Burgundy, Fort St. Louis, and additional defenses, such as at Fort St. John on Lake Ponchartrain.

had a large table covered with fine green cloth placed in the hall of the government house, on which the reports were laid. A lighted wax taper for melting the sealing wax was placed by the side of a new silver stand dish, which appeared to have been made for the occasion: the workmanship was well done, but the construction and form of the different parts was very whimsical. The sand-box was in the form of a drum, braced with fine silver wire, and ornamented with engravings, representing various implements of war. The vessel that contained the ink represented a bedded mortar, which could be elevated and depressed at pleasure, as occasion might require, and was likewise decorated with engravings; this device the Governor observed was in character, as the matter drawn from the mouth of the vessel frequently proved very destructive. The pounce-box was in the form of a glove or sphere, on which was engraven the equator, ecliptic, colures, tropicks [tropics], etc.

After a short dissertation upon this stand dish, the manufacturing of which I presume delayed for several days the execution of the instruments, the Governor and myself, seated ourselves at the table and signed the reports; they were then handed to our secretaries and attested.

During my residence at New Orleans, a number of curious documents fell into my hands, particularly a packet from Mr. Hutchins, containing a number of letters to his friends in London. One of the letters had enclosed two certificates for drawing his pay as a British officer, one of them bears date subsequent to the arrival of Governor Sargent in the Mississippi territory. They are in the following words.

> Major Anthony Hutchins makes oath, that he had not between the 24th day of December, 1797, and the 25th of June following, any place or employment of profit, civil or military, under his Majesty, besides his military allowance as a provincial officer.
>
> Sworn before me, the 2nd day of January, in the year of our Lord, 1799.
> (Signed) Anthony Hutchins.
> (Signed) Isaac Johnson.

The first paragraph of the letter which covered the foregoing certificates, directed the manner in which the money was to be disposed of, and is in the following words.

'Natchez, 25th January, 1799.
John Miller, Esq.

Dear Sir,

I send this only to enclose my certificate in hopes it may arrive safe, and as you will receive the amount, I will thank you to purchase two London state lottery tickets. One for my wife Ann Hutchins, and the other for my eight children, Samuel, John, Mary, Elizabeth, Nancy, Magdaline, Charlotte, and Celeste Hutchins, and have them recorded in their names, in the lottery office as formerly.'

It is presumed those documents are sufficiently conclusive to prove, that it would have been improper for me to have countenanced Mr. Hutchins's plan of placing himself at the head of the inhabitants of the district of Natchez, particularly during the existence of Mr. Blount's intrigues in favour of the British government. On the contrary, I considered it my duty, to oppose all foreign influence within the limits of the United States, and therefore, if possible by the aid of the friends of our government, in that territory, to destroy the influence, and destroy the power of Mr. Hutchins and his party in that country. The success after a contest of some months, may be said to have been complete: but a change of the administration of the United States, and an affected change of political opinions in Mr. Hutchins, and some of his party, had again brought them into consequence: thus circumstanced they can only be viewed in the light of political mercenaries, and as such ought to be guarded against.[44]

Here I wish to be indulged in an observation or two relative to the conduct of Winthrop Sargent, Esq. late Governor of the Mississippi territory. It appears that Mr. Sargent, either knowing or suspecting, that Mr. Hutchins was a military officer on the British establishment, refused to gratify him with any appointment whatever within his gift in that territory, and for which he open and candidly gave his reasons; which were so offensive to Mr. Hutchins, that he became his decided enemy, as he had before become mine, and in like manner declared his determination to effect his removal.

44. Ellicott is writing *The Journal* after his return to Philadelphia in 1800, and he is referring here to the election of Thomas Jefferson, Democratic Republican, in 1800, and the new administration, which took power in 1801. See Noble E. Cunningham Jr., *The Jeffersonians in Power: Party Operations, 1801–1807* (Chapel Hill: University of North Carolina Press, 1963).

From the conduct of Mr. Sargent on this occasion, it is evident, that let his political sentiments be what they might, his regard for the honour and reputation of his country was such, that he spurned the idea of giving an appointment to a person whose attachment to the government of the United States was suspected, though he well knew at the time, that a large share of his popularity would depend upon his acquiescence.

Although those documents concerning Mr. Hutchins, must be conclusive with every person who reads them, they are not more so, than those I saw relative to the plan already mentioned, for the effecting of which a number of our citizens received considerable sums of money from the Spanish government; the difference is, I am not yet at liberty to make the same use of them.

On the 1st of March, our vessel being completed, we proceeded down the canal to Bayou St. John's, and the next day gave the Governor and officers of government, an entertainment at the drawbridge. The day following we passed the fort at the mouth of the Bayou, or Creek, which was saluted by my escort, and returned by a discharge of artillery from the fort. The winds were very unfavourable, which delayed our arrival the end of the guide line on the Mobile, until the 17th in the evening.

At New Orleans, I was able to engage but two sailors, and they were both deserters from a British privateer, which lay some days off the mouth of the Mississippi. With those two sailors, who were completely illiterate, I undertook to navigate the vessel. Several masters of vessels offered their service, but the price they demanded was so high, that it was thought more economical to do it myself.

Here it appears proper to make an observation respecting the Pascagola, which is a large river, between Pearl river and the Mobile. It is navigable for small craft a considerable distance above the boundary, and from the report of some of my people who descended it, it is very deep, and falls with a number of small waters in a bay opposite to Horn Island: but the bay and mouth of the river, being full of shoals and oyster banks, it appears only adapted to the coasting trade.

CHAPTER VIII.

The author begins and completes a course of observations—opposition expected from the Creeks—writes to Col. Hawkins upon that subject—leaves the encampment and sails to Mobile point—some account of Mobile river and the town—arrives at Pensacola—joined by Col. Hawkins—interview with Governor Folch—treaty with the Indians at Miller's place—observes the transit of Mercury—makes a course of observations on the Coenecuh [Conecuh]—returns to Pensacola—difficulty with the Indians—account of the Coenecuh and city of Pensacola—leaves Pensacola and arrives at the end of the guide line on the Chattahocha [Chattahoochee]—makes a course of observations—treaty with the Indians—descends the river, to the mouth of Flint river—makes a course of observations—menaced by the Indians—retreats—interview with Mr. Bowles and a British officer—letter to Col. Hawkins on that subject—arrives at St. Mark's—account of the Chattahocha river, West Florida, and its importance to the United States.

The observatory being erected before we arrived at the end of the guide line on the Mobile, the instruments were up on the 18th of March, when a course of observations was begun, and completed on the 9th of April following: for the results and difficulties we met with in carrying the line over the Mobile swamp, see the Appendix.

We had now passed through the Chocktaw nation without interruption, but from the conduct of some Creeks or Muscogees, who had visited our camp, it was evident we should meet with difficulties from that nation.[45] I there-

45. On the cultural history of the Choctaws, see the study by Greg O'Brien, *Choctaws in a Revolutionary Age, 1750–1830* (Lincoln: University of Nebraska, 2005).

fore sent off on the 23rd of March, a special messenger to Col. [Benjamin] Hawkins, our principal agent of Indian affairs for the southern department, requesting an interview with him and some of the principal chiefs, at Pensacola about the 20th of April following, that we might fall upon some plan to ensure safety to our party, and success to the execution of our business.[46]

While at Natchez, I wrote several letters to Mr. Mitchell, one of the deputy agents for Indians affairs, enclosing others for Col. Hawkins, and though delivered to Mr. Mitchell, they were either suppressed by him, or intercepted afterwards, as they were never received by the Colonel. The former I am inclined to believe was the case, as Mr. Mitchell was evidently deceived, and impressed with unfavourable sentiments of me, either by the Commandant at Natchez [Guion], or some of the persons already mentioned, who visited him when in search of my letters. But in this there appears to have been a systematic plan, for although Col. Hawkins wrote a number of letters to me, not one ever came to hand, (until the return of the special messenger,) which effectually prevented our communication of sentiments during a very important and critical period.

A few days after our arrival at the end of the guide line on the Mobile, Mr. Gillespie was despatched up to Fort [St.] Stephens [located 67 miles north of Mobile], with an Hadly's sextant, to take the latitude of that place, and a sketch of the river, which he executed with his usual promptitude, the latitude is probably sufficiently accurate for geographical purposes.

The face of the country, and soil between the Pearl and Mobile rivers, are similar to those already described between the former and Thompson's Creek.

The 10th of April the instruments were taken down and packed up, and on the 11th I descended the river to the city of Mobile; on the same day a party was sent through the Mobile swamp, to extend a guide line to the Coenecuh, (commonly called the Escambia.)

The 13th I set sail for Mobile point, where I arrived in the evening, and came to an anchor; that night the wind shifted, which prevented the vessel from crossing the bar for several days; but the delay gave me an opportunity of determining the latitude with considerable precision.

46. For the role of Benjamin Hawkins with the southern tribes, see Florette Henri, *The Southern Indians and Benjamin Hawkins, 1796–1816* (Norman: University of Oklahoma Press, 1986).

The Mobile is a fine large river, and navigable some distance above the boundary for any vessel that can cross the bar into the bay. One square rigged vessel has been as high as Fort St. Stephens, in latitude 31°33′44″.

When the river is low, the tide ebbs and flows several miles above the line, and is sometimes observed as high as Fort St. Stephens [located at the falls of the Tombigbee] ; but when full there is but little, if any tide above the city of Mobile. It was in the latter state when I ascended it, and notwithstanding the current being constantly against us, and but little fair wind, we reached the place of our encampment north of the boundary in four days, my vessel was 38 or 40 tons burthen.

About six miles north of the boundary, the Tombeckby [Tombigbee] and Alabama rivers united, and after accompanying each other more than three miles, separate; the western branch from thence down to the bay, is called Mobile. The Alabama retains its name, until it joins some of its own waters, which had been separated from it for several miles, and then takes the name of Tensaw [Tensa], which it retains until it falls into the bay.

The easiest way from the Gulf of Mexico by water into the United States, is up those rivers, the navigation of each being equally good.

The upland on those rivers is of an inferior quality from their mouths up to the latitude of Fort St. Stephens, and produces little beside pitch-pine and wire-grass; but is said to become better as you ascend the rivers. The lands on those rivers have notwithstanding had a good character for fertility; but this has arisen from not discriminating between the upland which is generally unfit for cultivation, and the banks of the rivers, which are fertile in the extreme, and to which agriculture is almost wholly confined for a number of miles above the boundary. But those lands are subject to a great inconvenience from the inundations of the rivers.

Planting is not attempted in the spring, until the waters have subsided; and it sometimes happens, that inundations follow the first fall of the waters in the spring, and wholly destroy the previous labours of the planters. This was the case in May 1799, after the corn was two feet high: But this inconvenience is by no means so great as it would be in a more northerly latitude, there still remains summer sufficient to bring a crop of corn to full maturity.

The large swamp through which the rivers meander after their separation above the boundary is intersected in almost all directions by smaller water courses, which maintain a constant connection between the main branches;

such of them as were used by our people in passing, and repassing from one side to the other are laid down on the map.

At the mouth of the Mobile river, and on the west side, stands the town, or city of Mobile. The situation is handsome, and some of the houses are tolerably good, and for so small a place, the trade is considerable; but it is said to be unhealthy during the months of July, August, September and October.

The fort is of brick, and stands a short distance below the city; it is a well built regular work, and was taken from the British during our revolutionary war, by Don [Bernardo de] Galvez, who commanded the troops of his Catholic Majesty.[47] Since that time, it has been rebuilt, and put in a good state of defence.

From the traverse of the river between the boundary, and city, the latitude of the latter appears to be 30°36′30″ north, and longitude, 5^h 52′17″ west from the royal observatory at Greenwich.

The bay is extensive, and supposed to be about 9 leagues in length; but too shoal for the large shipping. The latitude of the bar, at the entrance into the bay from the Gulf of Mexico, I found by a mean two good observations to be 30°12′30″ north, and as the course of the bay is nearly north and south, the longitude must be nearly the same as that of the city.

On the 19th, in the morning, the wind serving we crossed the bar, and sailed for Pensacola, but the wind dying away, we did not reach the battery at the Cliffs [Fort San Fernando de las Barrancas], until after nine o'clock in the evening; where we were detained until ten o'clock the next day for examination. The wind serving in the afternoon, we sailed up to the city of Pensacola: but on the way, were obliged to come to a few minutes under the stern of a forty four gun frigate, and produce our passports. On my arrival at Pensacola, I found elegant, and convenient lodging provided for me, which I had reason to believe, were at the expense of the [English commercial] house of Panton, Laslie [Leslie], Forbes & Co., but on this subject I made no inquiry.[48]

The 25th of April, late in the evening, Col. Hawkins arrived, and the next morning about ten o'clock, we waited upon Governor [Vincente] Folch [y

47. See Caughey, *Bernardo de Galvez in Louisiana*; and Coker and Rea, eds. *Anglo-Spanish Confrontation.*

48. See J. Leitch Wright Jr., "The Queen's Redoubt Explosion in the Lives of William A. Bowles, John Miller and William Panton," in *Anglo-Spanish Confrontation,* ed. Coker and Rea, 177–193; and Coker, *Indian Traders of the Southeastern Spanish Borderlands.*

Juan], who in a short time informed us that he had that morning to give audience to two Seminole Chiefs; upon this we withdrew to Capt. Minor's quarters, which were within the Governor's inclosure, and a few minutes afterwards; saw two Indians go into the Governor's house: in less than an hour he joined us, and observed, that the Chiefs gave a strong talk against running the line, and they were Seminoles; the subject then passed over.

Col. Hawkins and myself, after some consultation, were of the opinion that the proper place to meet the Indians, who were then on their way, would be as nearly as we could judge on the Coenecuh [River], where the guide line would cross it: but the officers of his Catholic Majesty were of a different opinion, and proposed the city of Pensacola, which would give them considerable advantage over us in point of intrigue, at which they were habitually dexterous; and what was equally to be dreaded, the delay that might reasonably be expected from intoxication, in which the Indians always indulge themselves at treaties where liquor is to be had: after some conversation on this subject, the officers of his Catholic Majesty gave way, and on the 28th, in the afternoon, Col. Hawkins, Capt. Minor, and myself sailed to the head of the Bay. As soon as we came to anchor, Col. Hawkins went on the shore to meet the Mad Dog Chief, who was the speaker of the [Upper Creek] nation, and who we were informed had just arrived. The Chief immediately informed Col. Hawkins on his landing, that two Indians had just gone to the Tallesee's [Talapoosa] with bad talks from the Governor. The Col. assured him it was impossible, that the Indians alluded to were Seminoles, and had gone out to their nation. An Indian standing by, observed that the Mad Dog was right, that he himself saw the two Indians at Pensacola, and conversed with them, that he knew them to be Tallesees and that they had bad talks.

The Mad Dog wanted to come immediately on board, and give Capt. Minor, and myself the information; but Col. Hawkins prevailed upon him to defer it until the next morning, when he came on board each of our vessels, and gave the information separately: and proposed sending a runner after the two Indians immediately, but Capt. Minor conceived it to be wholly unnecessary, as he was confident the Indians in questions were Seminoles, and not Tallesees, in consequence of which no more notice was taken of it at that time.

On the 29th, we rode up the river to Miller's farm, which was the place agreed upon to meet the Indians. The day following they began to assemble, but from their slow movements in treaty business, it was not expected that we should finish our conferences with them in less than ten or twelve days. I therefore made the necessary preparation for observing the transit of Mercury, which was by calculation to happen on the 7th of May in the forenoon. The observation will be seen in the Appendix.

Two or three days before our public conferences took place with the Indians, the Mad Dog asked Col. Hawkins and myself, if we supposed that Governor Folch would attend the treaty: to which we answered in the affirmative. 'No replied the Mad Dog, he will not attend, he knows that I shall say to him about his crooked talks. His tongue is forked, and as you are here, he will be ashamed to show it. If he stands to what he has told us, you will be offended, and if he tells us that the line ought to be marked, he will contradict himself; but he will do neither, he will not come.'

On the 4th of May, we were joined by Col. Maxant, and several other Spanish officers. Col. Maxant represented Governor Folch, who was taken so unwell on his way to the treaty, that he thought proper to return back to Pensacola! So soon as the Mad Dog discovered that Governor Folch had returned to Pensacola, and was not going to attend the treaty, he called upon Col. Hawkins and myself, and with some degree of pleasantry said: 'well the Governor has not come, I told you so, a man with two tongues can only speak to one at a time.'

The 5th of May, Col. Hawkins and myself, had several conversations with the officers of his Catholic Majesty, in which the Colonel stated that many crooked talks had some time since been sent out among the Indians, that they had been taught to believe, that his Catholic Majesty had no desire to have the line determined or marked. Whether those talks had the sanction of the officers of his Catholic Majesty, or made by interested traders, he should not attempt to determine: but they had had their effect, and he saw but one way to remedy it, which was that Col. Maxant, (who represented Governor Folch) and Capt. Minor, the Commissioner, should in the most explicit manner, declare to the Indians, that it was not only the wish, but determination of his Catholic Majesty to have the boundary determined.

On the evening of the same day, the Mad Dog as speaker of the nation, informed us that the principal Chiefs had arrived, and would meet in council the next morning, and that it was expected by them the white people would open the conference.

The 6th about ten o'clock in the forenoon we met, and the business was opened by Col. Maxant, who went into an explanation of the late treaty between the United States and his Catholic Majesty, and ended by an unequivocal declaration, that it was the wish of his master the king of Spain, to have the treaty carried into effect, and the boundary determined and marked. Col. Maxant was followed by Capt. Minor, who explained in a forcible manner, the nature of our business: that the line we were tracing, was not a line of property, but of jurisdiction, a line between white people, and not intended in any way to affect the Indians in either their property, manners, customs or religion. These gentlemen went so fully and explicitly into the nature of the business, that Col. Hawkins, and myself, had little more to say, than to assent to what the others had advanced.

Col. Hawkins reminded the Chiefs of their treaty with him at Colerain, in the year 1796, and called upon them for a fulfillment of their stipulations, which closed the business on our part. The Chiefs after a short consultation among themselves, gave us to understand, that they intended to reply the next day.

On the 7th, the Chiefs by their speaker the Mad Dog, delivered their answer, which was short and satisfactory. They observed that many crooked talks had been sent into their country; but now they had seen and heard the representatives of both nations, and found that their talks were the same and straight. That being the case, they were perfectly satisfied, and with pleasure concurred in the determination of marking off the boundary, and would agreeably to their stipulation with Col. Hawkins at Colerain, send on their Chiefs and warriors, as an escort to the surveying parties.

The 8th we rode up to the Coenecuh, and fixed upon a proper place for encamping, and erecting an observatory, which was immediately begun and completed the next day, and the instruments set up. Our instruments, baggage and provision, had been taken up by water in boats and periagusas [pirogues], to the place of our encampment. The observations were completed on the 20th, and the instruments taken down and packed up on the 21st.

On the 22nd of May, Mr. Gillespie, the surveyor on behalf of the United States, began the guide line from the Coenecuh to the Chattahocha, being escorted by the military of the United States, and his Catholic Majesty, together with two Chiefs and twenty warriors of the Creek nation, agreeably to the stipulation already mentioned, made at Colerain.[49]

In the afternoon of the 23rd, Col. Hawkins and myself set off down the river in a canoe for Pensacola, where we arrived on the 26th in the morning,

49. For a look at the culture of the Creek Nation, see Ethridge, *Creek Country*.

when I took possession of the lodgings prepared for me before my first arrival. As it appeared probable, that if any intrigues should be set on foot with the Indians, to induce them to oppose the completion of the line, they would originate at Pensacola, we therefore concluded to stay a few days, to watch equally the conduct of the Governor and the Indians, till the surveyor should have proceeded so far on its way to the Chattahocha, that there would be no doubt of his reaching that river.

We had been but a few days at Pensacola, when Col. Hawkins was informed by a confidential Indian, that a large body of the Upper Creeks were on their way to that place in consequence of an invitation from Governor Folch, that the talks were crooked, and the line would be stopped. This information the Colonel immediately communicated to me, and after a little consideration we thought it would be most prudent to be wholly silent on the subject, and affect the most perfect confidence in the candour of the Governor. This intelligence was followed by information from the surveying party, that they had been menaced by a body of Indians. By this time it was generally known, that a large number of Creeks were on their way, and might be daily expected. The Governor affected to be entirely ignorant, either of the reason of their coming, or of their business. They at length arrived, and encamped about three miles from the city, when the Governor immediately left the place. The Indians were nevertheless determined to have an interview with him, as they had come in consequence of his invitation. Col. Hawkins and myself being desirous to bring the curious business to a close, which now began to wear a disagreeable aspect, [we] requested Capt. Minor to write immediately to the Governor, and hasten his return. On his return, he requested Col. Hawkins and myself to be present the next day, at ten o'clock in the forenoon at his house, when he was to receive a visit of ceremony from the Chiefs; but the cause of their coming being unknown to us, we declined attending.

The governor at this time certainly found himself much embarrassed between us and the Indians. To deliver talks, and issue presents to those who resided a great distance north of the boundary, and within the United States, would have a strange appearance to us, and to send them away without any, after an invitation, would give them great offence. The Governor at length informed them, that as they were north of the boundary, and that being now marked, they must look to the United States for presents, and deliver their talks to Col. Hawkins. In consequence of the direction which the business now unexpectedly took, we were in our turn called upon for presents. It was in vain to tell them that they came without our invitation, and therefore not entitled to any, some must be given, and a compliance was a matter of course. After receiving from us to the amount of two or three hundred dollars, they left the place apparently well satisfied. During those transactions, which were not closed until after the 20th of June, a number of letters passed between Col. Hawkins and Governor Folch on the subject of Indian affairs: in this correspondence the Colonel manifested that firmness, caution and candour, for which he has been so justly esteemed. Copies of all those letters were handed to me and transmitted to the secretary of state; but as they properly belong to Col. Hawkins's department, I shall forebear to publish them.

Before Col. Hawkins and myself left Pensacola, we called upon Governor Folch, and requested him to send an agent immediately among the Lower Creeks and Seminoles, to quiet their minds and explain to them the nature of our business. To which the Governor assented without hesitation, and assured us in the most unequivocal manner, that it would be done immediately: but notwithstanding this solemn declaration, it was not done.

Immediately after dismissing the Indians, Colonel Hawkins set out for his station among the Upper Creeks, by the way of Mobile, and on the 24th I proceeded down the bay to the bar; but the wind failing, we were detained until the 2nd of July. These calms are very common in that season along the coast. While laying at anchor I drew up the following short account of the Coenecuh river, and the bay and city of Pensacola.

The Coenecuh has generally, though erroneously been called the Scambia and Escambia, which is the name of a much smaller stream that falls into it from the west, a short distance above Miller's place, where we had our conference with the Indians, and observed the transit of Mercury.

The banks of the Coenecuh during a large portion of the spring, are inundated for many miles above the line, down to Pensacola bay, with very few exceptions. The upland is poor as far up the river we saw it, but it was said to be tolerably good about the head branches.

The river is navigable for small crafts at considerable distance above the boundary. All our tents, stores, instruments, etc. were taken up to our camp by water. The tide ebbs and flows but a few miles up the river.

The Coenecuh falls into the head of Pensacola bay, which is a beautiful body of water, well stored with a variety of fine fish, crabs, and oysters, and is justly considered one of the best harbours on the whole coast. Vessels not drawing more than twenty-one feet water, may cross the bar at all times with safety.

Pensacola stands on the west side of the bay, the situation is delightful, and the place remarkably healthy, but the water is shoal in the front of the city. It was the capital of West Florida, while that province was in the possession of his Britannic Majesty, when it made a very respectable appearance; but since the conquest [1781] of the colony by the Spaniards under Don [Bernardo de] Galvez, it has been on the decline.

The old fortification stood on some sand hills back of the city, and too distant to yield it any substantial protection; notwithstanding this circumstance, the Spaniards never once attempted to molest the inhabitants, or injure the buildings, during the siege of the forts, which lasted two months. The garrison made a gallant defence, and the surrender was hastened by one of the magazines accidentally blowing up. During the whole siege, as well as after the surrender Don Galvez conducted himself both as a man of courage and humanity. Mr. [William Augustus] Bowles, (commonly called General Bowles); Mr. Philip Key of the state of Maryland, and several other Americans of distinction, were at that time officers under Gen. Campbell who commanded the troops of his Britannic Majesty.

The trade of Pensacola, is at this time principally carried on by the house of Panton, Laslie [Leslie], Forbes and Company. Mr. Panton resides at Pensacola, and Mr. [John] Forbes at Mobile, where they live in an elegant stile, highly esteemed for their great hospitality and politeness by all classes of people.

From a number of good observations, the latitude of Pensacola appears to be about 30°23′43″ north, and the longitude by our measurement from the Mississippi and traverse of the Coenecuh river is about 87°14′15″ west from Greenwich: but from the observations of Sir John Lindsay and Dr. Loirimer 87°40′: it may lie between the two, but I suspect much nearer to the former. The latitude of the bar at the entrance into the bay, is about 30°18′ north, and the longitude from our measurement and traverse 87°17′ west from Greenwich. This harbor as well as all the others on the coast of East and West Florida, is rendered much less valuable on account of the worms. They are so numerous in this bay, that a vessel's bottom has been known to ruined in two months: and it is absolutely necessary for all vessels not copper bottomed laying in the harbor to be hove down, cleaned and payed [covered with pitch], every five or six weeks.

The entrance into the bay is defended by a small fort on the west end of St. Rose's island, and a battery on the main land nearly opposite to it.

July 2[d], a very light breeze serving, we crossed the bar about eleven o'clock in the forenoon and steered for cape Blass, which we made early in the morning of the 4th, when the breeze which had continued to very light died away, and we came to an anchor. From the time we weighed anchor on the 2[d], until we reached cape Blass, the breeze was so regular that we never once found occasion to shift a single sail, and the sea almost as smooth as a mill pond. On this passage we were a few hours out of sight of land, a phenomenon which I observed for the first time.

Shortly after we came to anchor at cape Blass, we were enveloped by a thick fog, or mist, which was followed by a gale of wind. We immediately got under way, and stood for the entrance into St. George's Sound. The gale continued to increase until near noon, but we received no other damage than springing our bowsprit. At one o'clock we entered the sound, and come to an anchor. The day following we began to search along the north side of the sound for the mouth of the Chattahocha river; but found the cost so intersected, and cut up by numerous water courses, nearly of the same magnitude, that the true channel was not ascertained until the 13th, when a fair wind serving, we ascended about thirty miles, and afterwards warped for two days more; but the labour being very hard, the weather extremely hot, and our progress slow, I left the vessel and proceeded up the river in an open canoe, having our apparatus with me.

My journey up the river was disagreeable and painful, being blistered by the rhus radicans, (poison vine,) from head to feet. This aptitude to be disordered by this poisonous vegitable [poison ivy] I have been subject to from my infancy, and have generally been confined in consequence of it, at least one week every summer since. The evaporation from the dew from this plant in the morning; falling upon me, is sufficient to produce the effect. The irritation and heat of this complaint was frequently so excessive, that I had to plunge into the river many times in the day, and lay whole hours in it during the night, which was the only relief I could find; medical aid had at all times proved ineffectual to relieve me.

On the 21st, I arrived at an Indian village, where I remained until the 23rd, when the horses arrived, and after they had rested a few hours, I set off for the camp, which I reached on the 25th in the evening in the midst of a heavy shower of rain. The canoe in which I ascended the river to the Indian village, arrived at our camp with the instruments, tents, etc. on the 26th. The observatory was finished on the 27th, and the instruments unpacked and set up;

but the rain continued until the 30th, and prevented any observations being made until that day; the course was completed on the 19th of August.

From an apprehension that difficulties might possibly arise with the Indians at that position, Col. Hawkins had written to Mr. Timothy Barnard, one of our deputy agents, to meet us on that river and explain the nature of our business to the Indians, if they should appear discontented with our proceedings.

Mr. Barnard arrived five days before me, and as it was evident we could proceed no further without an explanation, Mr. Barnard had taken measures for assembling the chiefs. But previous to our arrival, the Indians on the east side of the river, had assembled in considerable numbers to stop, and plunder the Surveyor; but his movements were so rapid, that he had arrived at the river, and was well posted before their main body had crossed, and their spies had fortunately been looking out too low.

On the 15th of August the Indians were assembled, and we proceeded to the conference: a number of speeches were delivered on both sides, and the business appeared to terminate as favourably as could be expected, and the Indians declared themselves perfectly satisfied: but I nevertheless had my doubts of their sincerity, from the depredations they were constantly committing upon our horses, which began upon the Coenecuh, and had continued ever since; And added to their insolence, from their stealing every article in our camp they could lay their hands on. Those doubts I frequently communicated to our agent Mr. Barnard, but who on his part had none.

It will be but justice to observe, that this disposition to plunder and impede our business, was by no means general: great part of the upper Creeks, the tame king, and his people excepted, behaved uniformly well, and upon a report reaching some of their towns, that the Seminoles, and Euphales [a band of the Lower Creeks] intended abusing us, and plundering our camp, a large body of them flew to our assistance, and offered to protect us through the country to the source of the St. Mary's; but their services were objected to by this Catholic Majesty's commissioner on account of the expense, who proposed as a substitute, to send to Fort Wilkinson for a troop of American horse which we were informed was stationed at that post: but this measure would probably have been attended with bad consequences, because it would have given the Indians reason to believe, that it was not his Catholic Majesty, as they had been frequently though improperly informed, but the United States alone, that wanted the boundary ascertained, and were determined to carry it through by force. Resistance would certainly have followed. But it would have been improper on another account; because it would be imposing an additional burden upon the United States, beyond that borne by his Catholic Majesty, which ought not to be expected, especially when it is recollected that the impediments which were thrown in the way by the officers of his Catholic Majesty, delayed the commencement of the business for more than a year; during which time the expenses of the United States were nearly the same as if the work had been progressing: And again, a troop of Spanish horse were stationed at Pensacola, which might much more readily have come to our assistance; but this, for some reason could not be agreed to by the Spanish commissioner.

The friendly Indians, who had come to our protection, and offered their assistance through the country to the source of the St. Mary's were thanked for the conduct, and good intention, and dismissed.

About this time, we received an account of the death [July 18, 1799, from yellow fever] of the late Governor General Gayoso, which I then, and yet consider a great loss to our western citizens concerned in the Mississippi trade, to whom he paid particular attention, and who frequently partook of that hospitality, for which he was so highly esteemed. As the Governor of an arbitrary monarch, he was certainly entitled to great merit, and it appeared in an eminent degree to be his pride, to render the situation of those over whom he was appointed to preside, as easy, and comfortable as possible, and in a particular manner directed his attention to the improvement of the country, by opening roads which he considered the arteries of commerce. He was educated in Great Britain, and retained in a considerable degree the manners, and customs, of that nation until his death, especially in his style of living. In his conversation he was easy and affable, and his politeness was of that superior cast, which showed it to be the effect of early habit, rather than an accomplishment merely intended to render him agreeable. His passions were naturally so strong, and his temper so remarkably quick, that they sometimes hurried him into difficulties, from which he was not easily extricated. It was frequently remarked of him as a singularity, that he was neither concerned in traffick, nor in the habit of taking doceurs [bribes], as was too frequently the case with other officers of his Catholic Majesty in that country. He was fond of show and parade, which he indulged to the great injury of his fortune, and not a little to his reputation as a paymaster. This fondness for parade showed itself in all his transactions, but in nothing more than ordinary business of his government, to which, method and system, were too generally sacrificed. In his domestic character he merited imitation: he was a tender husband, an affectionate parent, and a good master. In his correspondence with me relative to the late treaty, it is presumed he was governed by his instructions, and therefore no conclusion ought to be drawn from that discussion to his disadvantage as a man, and gentleman.

One or two days before we left our position on the Chattahocha for the mouth of the Flint river, Mr. Burgess, who had lately been one of our deputy agents, and interpreters, and who had agreeably to the Creek custom intermarried with several of their females, who then lived with him, informed me confidentially, that a plan was laid to plunder us on our way to the St. Mary's, and requested me to write to Col. Hawkins, to join us at the mouth of Flint river immediately, as his influence would affect our safety, if it was in the power of any man to do it. Upon taking leave of me he observed, that for certain reasons it would not then be proper to mention the information he had communicated; but to be careful to take such measures as would secure the safety of the party.

The night of the 22nd was wholly spent in writing and copying my despatches to the Department of State, and to Col. Hawkins, which were forwarded with all possible expedition by a runner to the Colonel.

On the 23^{d} Capt. Minor and myself, proceeded by water down to the mouth of the Flint river, and on the 25th began a course of observations, which will be seen in the Appendix.

In the beginning of September, Capt. Minor dismissed his military escort agreeably to instructions which he had received from the late Governor General Gayoso, as early as the 14th of May preceding; he also discharged great part of his labourers, and sent away a large portion of the baggage of his department, and the only valuable part of their apparatus. As soon as this was done, he became very importunate to set out for the St. Mary's. In one of our conversations upon that subject in the Contractors tent, he was told that work must be finished at that place before we left it, and which could not possibly be completed before the 14th, admitting the weather to be favourable. And moreover, that I was desirous of seeing Col. Hawkins before I ventured upon the journey, the success of which was, in my opinion, at best, very doubtful; and further, that my commissary Mr. Anderson had just reported, that our remaining number of horses were scarcely sufficient to transport the requisite stores, baggage and apparatus to the St. Mary's: it therefore became necessary for our agent, if possible to obtain some security from the Indians, for the remainder. But that his (Capt. Minor's,) situation was very different, his military escort was dismissed, almost the whole of his labourers discharged, and great part of the baggage, and apparatus sent away; he would therefore have horses sufficient for his purpose, if some should to be stolen on the way. To which he answered, 'I suppose you will be angry but I must now tell you that those men of yours are no longer necessary.' This expression was certainly a hasty one, and made without reference to the subsequent part of the work; for there yet remained more than 150 miles of line, for the surveyors to trace, correct, and mark, by erecting a large mound of earth at the end of each mile (stones not being had,) which would have required our whole joint force, if he had not discharged one of his labourers, to accomplish in any reasonable time. It is true, that there remained but one point more for the commissioners to determine, which was the source of the St. Mary's. The Surveyor on behalf of the United States, (Mr. Gillespie,) was gone back to the Coenecuh for the purpose of correcting the guide line; if we went on before his return, he and his party might be plundered and abused by the Indians, and would have no place to retreat to nearer than Pensacola, or St. Mary's. Capt. Minor was told, that my situation was similar to that of passed pawn in the game of chess, and that it did not appear prudent to make another move, until supported by a piece, and if Col. Hawkins upon his arrival should be of opinion that appearances were favourable, I would immediately proceed, but not otherwise.

On the 9th Mr. Burgess paid us a visit. After dinner he took me into the observatory, and asked this question, 'Did you write to Col. Hawkins while at the Upper Camp agreeably to my recommendation?' To which he was answered in the affirmative. 'You have not,' says he, 'written as pointedly as was necessary, he would have been here before this: you must write to him immediately, and procure support from the Upper Creeks, which may be had, or you will positively be plundered on your way to St. Mary's; you may think me a fool, but mark the end.' He was told that the letter to Col. Hawkins was sufficiently pointed, and if he was well, there could be no doubt of his joining us in a few days.

The 14th of September Col. Hawkins joined us, and being of the opinion that every thing relative to the Indians was in a good train, and that might go on with safety, my objections were then immediately withdrawn, and I began the arrangement of the astronomic journal, that it might be understood, if any accident should happen to me, and it be preserved. The commissaries of both parties, were directed to have every thing ready for our moving on the 20th.

Early in the morning of the 17th, we received a message from Indian Willy, (a person of property,) who resides on the Chattahocha, a few miles above the mouth of Flint river, to the following effect: 'Gentlemen, I have sent my Negro, to inform you that about twenty Indians lay near my place last night, they intend mischief; many more are behind: they say they are Chocktaws; but this is not true: be on your guard, and remember I have nothing to do with it: my Negro goes at midnight.' Although this information was not slighted, it was not pointedly attended to. About two o'clock in the afternoon, some Indians belonging to our escort, were sent over the river to make discoveries; but returned in two or three hours, without making any; but at sun down, our doubts were at an end: we then received positive intelligence, that a number of strange Indians had just crossed the river; upon which my escort was immediately called to arms, and my escort and my son and labourers, who were armed with rifles joined it. The Big Lieutenant, (a Creek Chief,) who commanded our Indian escort, was directed to go, halt the strangers, demand their business, and give us immediate information. They halted but for a short time, and declared 'their object was to plunder the camp, scatter the people, and let them go home what way they pleased.' They then proceeded to within about two hundred yards of our camp, when they were again halted by our Indian escort, and interpreter; but still persisting in their determination to take and plunder our camp, we had only to choose between submission and resistance; the latter was immediately resolved upon, and Col. Hawkins requested Capt. Bowyer, (who commanded my escort,) to arrange his men in the best manner he could: the labourers, with their rifles, being stationed on the flanks. This was done with great expedition, and the party marched within about twenty yards of the hostile Indians, and was then halted by Col. Hawkins, who stepped forward, and addressed the strangers, who still persisted in their determination to plunder the camp, and told us, that if they met with no opposition no blood should be shed; but vengeance would follow resistance. The Colonel with great firmness replied, 'that we were willing to put it to issue, and if any one of their party should attempt to remove any one article in the camp, he should instantly be put to death, and if the party presumed to march one step nearer, it should instantly be fired on.' This parlaying continued until about ten o'clock in the evening, when the Indians became more cool, and agreed to remain quiet until the next morning, when they would hold a talk; but at the same time, gave us to understand, that they were determined to carry their design of plunder into effect; which they should be able to accomplish with ease, as their strength was hourly increasing. Upon the assurance of their remaining quiet until the next day, our armed party marched back, and guarded the camp until morning.

About three o'clock in the afternoon of the arrival of the Indians, Capt. Minor's riding horse, with another very valuable one, were stolen from within two hundred yards of the camp: upon receiving this intelligence, orders were immediately given to bring all those belonging to our party into the camp, and secure them, when we found that eight or ten were already missing. The Spanish horses were also collected, and put into a pen made for that purpose.

The hostile Indians kept moving about in small parties the whole night, and sometimes came within gun-shot of our tents. They threw down the contractor's bullock pen, and let his cattle out, and opened the pen which contained the Spanish horses, and haltered four of them; but were driven off with but two: Three of the horses belonging to my party broke loose, and ran without the camp, and though every exertion was made to bring them back, they were mounted by some Indians, who rode off on them. In this manner the night was spent.

From suspicion that we should meet with some difficulties at that place, I had detained a small Spanish schooner, which was in the employ of our commissary Mr. Anderson; the United States schooner which I had fitted up at New Orleans, was too large to ascend the river with ease, I had ordered her to Apalachy (St. Marks,) a few days before, and to wait further orders. The principal part of the loading of the small schooner, I had very fortunately taken to our camp a short time previous to the alarm. The fate of the vessel we did not learn until early in the morning of the 18th, when we were informed that she had been plundered about midnight; that the sails were cut to pieces, and the running rigging carried away. Upon receiving this information, my son with two of the labourers, armed with rifles, went to repossess her: on their way they saw a small party of armed Indians, who fled on their approach; as they drew near the vessel, they discovered three Indians onboard of her; seeing that their numbers were equal, they gave whoop, and sprang forward, on which the Indians jumped on shore, and fled with precipitation into the woods. The master and other persons on board, had been robbed of all their clothing, even their handkerchiefs from their necks, and heads, together with their bedding. The public property was of no great value: twelve or fifteen guns which wanted repairing, a case of claret, a small quantity of brandy, a chest of axes with other tools, and a few blankets were the principal articles taken.

We waited impatiently until nine o'clock in the morning, but heard nothing of any Indians coming to the conference they had proposed the preceding evening. We called upon the Big Lieutenant, and asked his opinion of the situation; he answered 'that it was far from being good: the Indians on the river about us had taken the bad talks of the strangers: that he had no dependence, but on his own, and our people; with them he thought he could take us safely to St. Marks, (Apalachy).' Upon receiving this information, Capt. Minor and myself thought it best to retreat. It was agreed that he should proceed by land to Colerain, if not followed the two first days by the Indians; but in case he were, to proceed to St. Marks, and wait for the vessel to carry him, and his party, round Cape Florida to St. Mary's: the vessel he had heretofore used was discharged; (I believe) by an order from the late Governor Gayoso. To render his journey as safe as possible, I sent Capt. Bowyer and all my escort, a corporal and three privates excepted with him.

I went on board the small schooner myself; in which was put the apparatus, with all our provision, and baggage, except what would be wanted by the party who went by land, determined to force my way down the river if opposed: my armed party consisted of sixteen persons, twelve of whom were expert rifle men.

Before I left the camp, I wrote a letter to our surveyor Mr. Gillespie, to be forwarded to him by Col. Hawkins, who proposed remaining on Flint river a few days, to endeavor to give a more favourable turn to the disposition of the Indians. A message was likewise sent to the two sailors on board the United States schooner at St. Marks, to meet me with all possible expedition in St. George's sound. The party who went by land with the remaining pack horses set off at four o'clock in the afternoon, and the schooner was under weigh [way] at five; we manned, eight oars and relieved the hands every hour.

Immediately after we left the shore, it began to rain, but we soon made such a covering with our tents, the cut sails, and some oil cloth, that our people and their arms were kept dry. We continued down the river until after dark, when we stopped for fear of injuring the vessel against logs. The next morning, although it was raining, the moon gave us so much light, as to enable us to proceed, and about eleven o'clock in the forenoon of the 19th, we passed the lowest Indian village on the river. The rain continued without intermission, and so heavy that it would have been impossible for the Indians to attack us with success in open canoes, and they have no other.

On the 21st of September, we came near to St. George's sound, where we halted to repair the rigging of the vessel. This ended this disagreeable business. It was alarming, because we had savages to deal with. To the American party, (who a few days before had been declared useless,) both camps were indebted for their safety, and by them public and private property, to a considerable amount, together with all the papers, observations, and other documents, relative to the boundary were preserved. The Spanish party was too small to make even the shew of resistence, those few however behaved with great firmness. Here it will be justice to add, that it is my opinion, the Spanish commissioner, (Capt. Minor,) was as much in the dark respecting the conduct of the Indians as we were, and was wholly regulated in his conduct by his instructions, and that if anything improper has been done by other officers of his Catholic Majesty, it was wholly without his knowledge.

The following statement of facts, may perhaps satisfy the reader on this subject.

First. In the month of May preceding, Capt. Minor received instructions from Governor Gayoso, to dismiss his escort on his arrival at the Chattahocha, and return himself to Pensacola, and wait there until he could be furnished with a passport from the Bahama Islands, that he might be enabled to go with safety round Cape Florida to St. Mary's. That their surveyor should go by land from the mouth of Flint river, and carry on a line east until it intersected the Apalachy, after which he should proceed to the source of the St. Mary's, and when the geographical position of that point was determined, the work might be considered as complete. These instructions were shown separately to Col. Hawkins and myself; what were the Colonel's comments on them I know not; but my own opinion of them was given very freely: They were considered as calculated for the purpose of delay: The directions to the surveyor, to carry on the line east from the mouth of Flint river, until it intersected the Apalachy, was not only unconnected with our business, but embraced an absurdity; because a line drawn east from the mouth of Flint river would pass many miles north of the source of the Apalachy. The orders respecting the military escort were certainly improper: it was a subject submitted to the commissioners of the treaty.

Captain Minor was not pleased with the instructions, and immediately by letter to Governor Gayoso proposed some alterations, who dispensed with his returning to Pensacola, and waiting for a Bahama passport, and with the east line to be carried on to the Apalachy.

Secondly. The two Indians, whom Governor Folch assured Col. Hawkins and myself were Seminoles, were Upper Creeks, from the Tallisee town, and brothers-in-law to his interpreter Antonio.

Thirdly. That instead of going to the Seminoles, they returned home to their town, of which the Tame King was Chief.

Fourthly. A few days after their return home, the Tallisee; or Tame King, with about two hundred of his people including the two already mentioned proceeded to Pensacola, and menaced our surveyor and his party on their way. An account of their proceedings at Pensacola has already been detailed.

Fifthly. The party who came down to plunder us, were a part of those who came to Pensacola with the Tame King.

Sixthly. Governor Folch promised Col. Hawkins and myself, when in Pensacola, that he would immediately send agents among the Indians, particularly the Seminoles, to quiet their minds, which was never done!

Seventhly. Had the officers of his Catholic Majesty been careful to have agents among the Indians, to cooperate with ours, in keeping up a good understanding, and allaying their fears and suspicions, our business would have been completed in all its parts.

On the 23^{d} in the forenoon, my two sailors arrived in an open boat. The United States schooner, could not be got out of the harbour, on account of a head wind. On their landing, they handed me two letters, over the superscription of one was written these words, 'On his Britannic Majesty's Service.' The following is a true copy of it.

Fox Point, September 22nd, 1799.

Sir,

I beg leave to make known to you, that I am at present on a small island on this coast, which is well known to the bearers, with the crew of his Britannic Majesty's schooner, Fox, late under my command, but which was unfortunately wrecked five days since on this coast. As there is no possibility of saving the schooner, I trust sir, your humanity will induce you to stop here, and devise with me some means of removing those unfortunate men, who have nothing more than some provisions saved from the wreck to exist on; the island producing nothing, on the contrary for two days, during the late gale, the sea made a breach over it, so that for those two days, we were with nearly two feet water on the ground.

Understanding that you have been driven by the Indians from the country where you were employed, I beg leave to inform you, that General Bowles, the Chief of the Creek nation is with me, he expresses his wishes to see you much, as he thinks your unfortunate differences may be settled: he has no force here, therefore you may be assured no treachery can be intended, as I shall consider you under my protection and use the force under my command to the utmost for your security, which is not inconsiderable, as I have been enabled to save my arms, ammunition, etc. With the most anxious wishes of seeing you soon,

I am your most humble servant,
(Signed) James Wooldridge,
Lieutenant in the Royal Navy.

Andrew Ellicott, Esq.

The other letter was in the following words.

Fox Point, September 22nd, 1799.

Sir,

I am now at the mouth of this river on my return from Spain by the way of London, and the West Indies, in order once more to join my nation the Creeks. The vessel that brought me here, was four days since unfortunately run on shore at the entrance of the bay, but having saved all my effects, with my boat, should have proceeded into the country, until hearing of your being near, I determined to stay, and wish much to see you. Although we may differ in politics, yet as gentlemen we may associate, and be friends, at least we may be civil to each other; I pledge my honour to be so to you and rely on yours,

I have the honour to be,
Your obedient servant,
(Signed) Wm. A. Bowles.[50]

Andrew Ellicott, Esq.

50. William Augustus Bowles (1763–1805) was born in the United States, served as a young British officer at Pensacola during the American Revolution, and then went to the West Indies,

It had been for some weeks previous to this time, rumored among the Indians, that Mr. Bowles was on his way to join them; but whether they had this information from himself while in the West Indies, or whether it had been communicated to them in consequence of some injudicious publications which appeared in several of our news papers is uncertain; but it evidently had a bad effect.[51]

The line of conduct which was thought proper to pursue on this sudden and unexpected incident, will be found minutely detailed, in the following letter to our principal agent Col. Hawkins.

Apalachy, October 9th, 1799.

Dear Sir,

On the 23r of last month, at the mouth of the Chattahocha, where my people were repairing the rigging of the vessel, which had been cut to pieces by the Indians on the morning of the 18th preceding, the

---

from where he returned to Florida and lived among the Lower Creeks. In a famous episode, he and his Indian allies captured the fort and storehouse of the Panton firm at Saint-Marc-des Apalaches in 1792, which therefore made him an enemy of Spain. Operating as an agent of the British government and a self-styled "chief" of the Creek and Cherokee Indians, he later organized his own "nation" in Florida. See the account authored by Bowles himself, *Authentic Memoirs of William Augustus Bowles, Esquire, Ambassador from the United Nations of Creeks, Cherokees to the Court of London* (London: R. Foulder, 1791). See also Elisha Douglass, "The Adventurer Bowles," *William and Mary Quarterly,* 2nd ser., 6 (1949): 3–23; Lawrence Kinnaird, "The Significance of William Augustus Bowles' Seizure of Panton's Apalachee Store in 1792," *Florida Historical Quarterly* 9 (1931): 156–92; and Gilbert Din's excellent overview of the relationship between Bowles and Spain, *War on the Gulf Coast: The Spanish Fight against William Augustus Bowles* (Gainesville: University Press of Florida, 2012).

51. In a discussion of the whole issue of trading with the Indians, James Pitot included an analysis of Bowles: "It was during the military government of [Marques de] Casa Calvo that Bowles—European [American] by birth, Indian by profession, who had escaped from the pursuit of the Spaniards and was finally employed by England—arrived on the Florida coast [1799] with weapons, gifts, and munitions in order to induce his Indian brothers to declare war against the Spaniards. He had been up against delays and adversities which had led him to Jamaica. There he had gathered together a kind of general staff. But much greater misfortunes awaited Bowles on the shore, and a shipwreck cast him on the beach with all his followers, after he had jettisoned the most essential items in his armament" (Pitot, *Observations,* 20). See also the discussion of events surrounding Bowles's capture by the Spanish and their Indian mercenaries in 1803 provided by the French colonial prefect for Louisiana in Laussat, *Memoires of My Life,* 36–39.

enclosed letters Nos. 1 and 2, were put into my hands by my sailors, who had come from Apalachy in an open boat to meet me; the United States schooner could not be got out of that harbor on account of a head wind. On their passage, they were brought to at the east end of St. George's island, by some people in distress. Upon going ashore, they were critically examined by one of the gentlemen respecting the employ they were in, and where they were bound; to which they gave satisfactory answers. They were then informed that the people they saw there, were the officers and crew of his Britannic Majesty's armed schooner the Fox, which had been wrecked at that place five days before, and that Gen. Bowles, and his suit were among them, and requested the sailors to be the bearers of some letters to me, to which they consented. On receiving those letters which was in the forenoon, I did not decide in what manner I should act until sometime in the afternoon, when I concluded to go immediately in the open boat to those unfortunate people, and leave the schooner, (which was in the employ of our commissary Mr. Anderson,) to have her rigging repaired. At five o'clock in the afternoon I set out, and proceeded along the sound until one o'clock in the morning of the 24th, when I wrapped myself up in my cloak, and slept on one of the benches; the men being much fatigued with rowing, likewise went to sleep. At day break they took in the anchor, and a light breeze serving, we arrived at the east end of the island about ten o'clock, A.M. where I met with the unfortunate crew, and after receiving an account of their misfortunes from the commanding officer, he was informed that their situation had been taken into consideration, and my mind was made up upon it. That the country which I had the honour to serve, and of which I was a native, had early resolved to observe a strict neutrality between the present belligerent powers in Europe. This resolution I thought a wise one, and could not therefore on any occasion, consent to any one act which might be construed into a deviation from that principle. That the officers and crew were certainly in their enemies [Spain] country, and came into it with hostile views; an attempt therefore on my part to extricate them, might be viewed by the Spanish government, as a deviation from that line of conduct, we had determined to observe.[52] From that view of the

52. Ellicott is correct here. The British schooner *Fox* was transporting Bowles and his associates secretly into Spanish territory for purposes of a covert operation in the war against Spain.

subject, which I thought a correct one, they were not to expect any other aid from me, than what was immediately connected with humanity; that when my commissary arrived, which would probably be the next day, I should direct him to furnish them as liberally as our circumstances would justify, and if I could be of any service to them in a negotiation with the officers of his Catholic Majesty, they might rely upon my interest and exertions in their favour. The officer, (Lieut. Wooldridge,) who appears to be a man of liberality, and good understanding, made no objection; it was therefore concluded that he admitted the justness of the principle. The next day the commissary arrived, and delivered to the Lieutenant 15 cwt. of flour, and three bags of rice; the crew were then on half allowance, great part of their provision being lost when they were wrecked.

I shall now proceed to take some notice of Mr. Bowles, (commonly called Gen. Bowles) who, with his suit, came in the vessel under the command of Lieut. Wooldridge. I had many conversations with Mr. Bowles, both of us being detained together eight days on the east end of St. George's island, by a violent gale of wind. He is certainly a man of enterprise, and address, added to considerable talents. He declared to me, that he was not taken by the Spaniards some years ago at Apalachy in the manner reported; but for political reasons, it was necessary to give it that appearance. That in 1794, it was proposed to him by the Minister of his Catholic Majesty, to receive a commission in that service, and return to his nation, and attack the United States; which he declined in a pointed manner, and was shortly after, and not until then confined. Soon after Mr. Pinckney arrived in Spain, he was informed by the Prince of Peace [Don Manuel de Godoy], that the American Minister was his enemy, and was again offered a commission which he declined to accept, though in confinement. That in the year 1797, if my recollection serves me, he was informed by the same Minister, that the Floridas were ceded to the Republic of France, and to be taken possession of as soon as convenient to the Republic, and was offered a commission in that service, which he also declined: And that immediately upon the late treaty between the United States, and his Catholic Majesty, being made public, he protested against it to the Minister of the court of Madrid, as interfering with the dignity of his people and nation, the Creeks, who were as free, and independent as any other nation in the universe. That the article by which the United States, and his Catholic Majesty, are bound to restrain the hostile attempts of the Indians within their respective territories, was an atrocious violation of the law of nations, and should never be submitted to whilst his people had a drop of blood to spill. He declared that he had arrested that part of the treaty at Madrid and that it must be done away by the executive of the United States. And further, that he had warned the court of Madrid against running the boundary, and expected from assurances given, that it had been suspended some months ago, and had also demanded in the name of his nation an immediate evacuation of the post of St. Mark's, which if not done immediately, he should fall upon measures to compel a compliance, and had he arrived in time, he should have arrested the Spanish commissioner, and his party. He likewise intends to seize Mr. Panton's property at Apalachy.

This is the substance of all the conversation I had with him interesting to our country. What credit may be due to his information, and what we have to fear from his threats, you are better able to judge of than myself: some Indians will probably be led away by him, and some temporary inconvenience experienced by the United States; but I cannot think that the nation generally, will risk its existence to gratify either his ambition or resentment. He speaks in the style of a King; "my nation,' and 'my people,' are his common expressions.

Whether he is supported, or countenanced by any foreign nation is a question that scarcely needs a conjecture. Certain it is, that on his arrival last spring at Barbadoes, he was treated with the utmost respect, and at Jamaica with singular attention, both by the Governor and British admiral on that station. This attention induced two young gentlemen to become his followers, the one a Scotchman by the name of Ferguson, and the other a Mr. Nuvelle, a French gentleman, now or lately a Captain in the Prince of Wales's regiment of colonial light dragoons; which is manifest from his commission. These gentlemen began to fear a deception, and suggested their doubts to me, and at the same time asked my advice respecting a relinquishment of their plan of proceeding into the nation. They were told, that their present situation was such, that I should forbear saying anything to them on that subject, but would furnish them with a line to you, and doubted not but in a few weeks they would be able to judge correctly for themselves.

Although the officers of his Catholic Majesty have certainly been very remiss, in obliterating the impressions which were made upon the minds of the Indians two years ago against running the boundary, I am nevertheless of opinion, that Mr. Bowles's plans, which are unquestionably hostile towards Spain, ought to be counteracted by every citizen of the United States.[53] It is not only a duty we owe to the supreme law of the land, but involves a point of national honour: a compliance with the most important of all contracts.

I have now my dear friend, given you as correct an account of my interview with Mr. Bowles, as comes within the power of my recollection, and feel a confidence of your taking such measures, as will preserve the honour of our country, and effectually thwart the views of a bold, daring, adventurer.

I cannot close this letter without observing in justice to Mr. Bowles, that he behaved on all occasions whilst with me in a polite and friendly manner, and generously furnished me with the necessary charts and directions, for sailing round Cape Florida, a matter of great importance to me, as I shall have to navigate our vessel myself.

Mr. Panton and Governor Folch, have been written to on the subject of Mr. Bowles's arrival in the nation, a copy of my letter to the latter, you will find inclosed.

53. The account presented by Bowles was part truth and part fiction. He claimed to have met with the Court of Spain, the Prince of Peace prime minister (Godoy), whose diplomatic priority was peace with the United States, not to start a war; Spain considered Bowles a pirate and agent of the British. His 1792 attack on Panton's store, which operated under an exclusive license for the Indian trade with the Spanish government, was an attack upon Spain. Bowles was captured after the attack, transported to New Orleans, and from there sent to Havana, then Spain, and finally to the Spanish fort in the Philippines. After his interview with Ellicott, Bowles would again venture into the Creek encampments with his plans, but they were foiled by a substantial reward offered to the Indians by the Spaniards for his arrest. At an Indian conference, they captured him (1803) and delivered him to the Spanish in New Orleans; Bowles died in a Spanish prison in Havana, Cuba, in 1805. See the account of his capture provided by Prefect Laussat in 1803 in his *Memoires of My Life,* trans. Agnes Josephine Pastwa, ed. with a foreword by Robert D. Bush (Baton Rouge: Louisiana State University Press, 2003), 36–39. According to James Pitot, who was on the scene of these events: "In fact, he [Bowles] gives proof of the Spanish government's weakness, and the meager influence and position which it has been able to maintain for itself among the Indians" (Pitot, *Observations,* 90). A comprehensive account of Bowles, and insight into the realities of Spain's administration, is provided by Din, *War on the Gulf Coast.*

I am my dear Sir,
Your sincere friend,
Andrew Ellicott.

Col. Benjamin Hawkins, principal
agent of Indians affairs for the
southern department.

My passage from the mouth of the Chattahocha, to St. Marks, (Apalachy,) where I arrived in the evening of the 7th of October, was truly disagreeable, being part of the time at sea, in an open boat, and detained thirteen days by violent easterly winds, on dry islands of sand, which was blown about like snow, filled our blankets, and fell in great quantities in what little victuals we had, which for seven days was bread and coffee, and what few fish we caught. On my arrival at St. Marks, I received a note from my friend Col. Hawkins, informing me that Mr. Gillespie our surveyor, had joined the party which had gone on by land. This intelligence eased my mind of many uncomfortable reflections.

Having now navigated the Chattahocha up to, and down from the boundary, and traversed the whole coast of West Florida, I shall proceed to give some account of that river, and the province.

The Chattahocha, (sometimes called the Apalachicola,), is a fine, large river, and navigable for boats, and gallies, which use oars a considerable distance north of the boundary. A sloop in the service of his Catholic Majesty's commissioner, and a small schooner in our employ, ascended up to the mouth of the Flint River, which falls into the Chattahocha about twenty-one miles below the parallel of 31°; but this was attended with some difficulty. The United States schooner ascended more than 30 miles; but for want of oars and hands, had to stop. From the mouth of the river up, for the distance of at least forty miles, the banks are very low, and with the exception of a few places, inundated whenever the water is moderately high. But as you ascend, the banks become more elevated; and some of them which may be called bottom land, are seldom overflown: these are remarkably rich, and extremely fertile, and are almost the only lands under cultivation by the Indians, who reside on the river.

A few miles below the mouth of Flint River, limestone begins to make its appearance, and extends very far up into the country; it is open and porous, and of a dirty bluish colour. On the east side of the mouth of Flint River, and for a considerable distance up it, large quantities of iron ore may be seen.

The up-land on the Chattahocha, and Flint Rivers, from the boundary southward is of an inferior quality, though much better than on some of the waters already mentioned.

The Chattahocha, communicates with St. George's Sound by three mouths or channels. The most eastern, is at present only navigable for canoes, and small boats on account of the lodged timber, and rafts. Our vessels ascended the most western one, which is at this time the main channel; but the navigation of this is troublesome for those not acquainted with it, though not on account of logs, and such impediments, but from its connection with lakes and swamps, by branches apparently as large as itself. We took two of them coming in from the westward, the first led us into a lake about three leagues in length, and a half in breadth. The other, a few miles from the main branch, was divided in such a manner, into small ones, that we soon discovered our mistake. The latitude of the mouth of the western branch is about 29°42′ N. and the longitude by a lunar observation 39′23″ west from the royal observatory at Greenwich.

St. George's Sound, is principally formed by three islands, between the most westerly one, and the main land, the channel is narrow, and shoal, and only fit for canoes; between this island and St. George's, which gives name to the sound, is a bar on which some bushes are growing. The coasting vessels pass between these islands. St. George's Island is supposed to be about six leagues in length, but in no place more than one wide. The distance from St. George's Island across the sound, is from one to two and an half leagues. The next island is not laid down in any of our charts: it is about two leagues in length, and two miles east of St. George's Island. The main channel into the sound, is near the west end of this third island. From this third island, to a fourth, which at low water sometimes joins the main land, the water is too shoal for any other than coasting vessels.

The latitude of the east end of St. George's island where the schooner Fox, (already mentioned,) was cast away, is 29°44′38″ N. and the longitude (by taking the result of a lunar observation made at the mouth of the Chattahocha as a correct point,) $5^h38'35''$ west from Greenwich. The sound is so full of oyster banks, and shoals, that it is difficult to navigate it, without a pilot.

The coast on the north side of the sound, is intersected and cut to pieces by such a variety of water courses, several of which have evidently at some former period been mouths of the river, that it is extremely difficult to find the true branches: of this we had sufficient evidence.

The up-land of West Florida, as it now bounded, is generally of a very inferior quality, except on the Mississippi, and is of but little value for either planting or farming. The river bottoms, or flats are all fertile; but too inconsiderable as to quantity, or, too low and marshy, to give much value to the province.

It may be observed, that no restrictions in this country, have been found so effective, as to prevent settlements being made where the land has been good. A conclusion may therefore be fairly drawn, that this province, which has been aided by France, Great Britain and Spain, each in her turn, and yet remain in a great degree unsettled, must be materially defective in point of soil.

It is true, that the cities or towns of Mobile, and Pensacola, have been flourishing places, but this was owing to causes not immediately dependent upon the soil. The latter was the seat of government while the province was held by Great Britain, and from the excellence of the harbor was much frequented by the shipping of that nation, and both places well situated for carrying on the Indian trade, which was at that time very considerable; but that trade having greatly decreased, from want of inhabitants, and necessary articles of exportation, those cities have declined also. Mobile is now beginning to recover, but this is owing to the settlements forming north of the boundary, on the Tombecby, and Alabama Rivers. Notwithstanding the favourable situation of those cities, they can never be of much consequence but from the settlement of the country north of the boundary, which has greatly the advantage in point of soil and climate.

Although West Florida is of but little importance, when considered alone, and unconnected with the country north of it, it is of immense consequence when viewed as possessing all the avenues of commerce to, and from a large productive country. A country extending north from the 31st degree of north latitude, to the sources of the Pearl, Pascagola, Tombecby, Alabama, Coenecuh, Chattahocha, and Flint Rivers, and at least 300 miles from east to west. The coast of this province abounds in live-oak and red cedar, in considerable abundance, fit for ship building, which is not to be met with north of the boundary.

From the safety of the coast of this province, added to the great number of harbours proper for coasting vessels; that of Pensacola into which a fleet may sail, and ride with safety, and that of St. Joseph's, into which vessels not drawing more than seventeen feet of water may sail at all times; it must be highly important in a commercial point of view, and if connected with the country north of it, capable of prescribing maritime regulations to the Gulf of Mexico.

In a political point of view, West Florida may be considered as an object of the greatest importance to a large division of the United States; because that power, which hold the avenues to commerce, may give a tone to the measures of another, should it be unfriendly to liberty, and public happiness.[54]

The population of West Florida is very inconsiderable. The principal settlement is on the Mississippi, between the boundary and the Iberville. On the north side of the Iberville, and the lakes, to the Gulf of Mexico there are a few scattering inhabitants. Thence along the coast, to Mobile Bay, there are a few more. There are likewise a few about the Bay. From the city of Mobile, up the Mobile and Tensaw Rivers, to the boundary, there may possibly be forty families. From Mobile Point to Pensacola Bay, there are no inhabitants, and not more than half a dozen farms on the Bay. From the head of the Bay, up the Coenecuh to the boundary, there are two plantations or farms. The population of the cities of Mobile, and Pensacola, does not exceed fifteen hundred inhabitants. From Pensacola Bay to St. Mark's, there are no inhabitants.

54. The Spanish-French secret treaty of 1800, which retroceded Louisiana from Spain back to France, and the growing demand for an American outlet to the Gulf of Mexico, had reached a fever pitch by 1801. The western states, Congress, and President Jefferson were adamant in their efforts to secure New Orleans, and/or "lands to the east" (Mobile or Pensacola) as ports for American commerce; it was generally believed in the United States by 1801, although incorrectly, that West Florida had been ceded by Spain to the French as well as the province of Louisiana. If Jefferson's French diplomacy failed, the mood of the country was for war, and Congress authorized the use of force and appropriations in 1803 with which to accomplish it. Fortunately, events in Paris resulted in the purchase of Louisiana and New Orleans in 1803, which resolved the "Mississippi Question" but not that of West Florida. See Robert D. Bush, *The Louisiana Purchase: A Global Context* (New York: Taylor and Francis), 14–35.

CHAPTER IX.

Occurrences at St. Marks—account of the adjoining country—the author sails from St. Marks—arrives at Cayo Anclote—thence at the Florida Keys—examines the Keys, and Reef to the Key Biscana [Biscayne]—sails into the Gulf Stream—a violent gale of wind—velocity of the Gulf Stream—another storm—one of the men dies—arrives at the west end of St. Simon's Island in the state of Georgia—theories of the Gulf Stream examined—arrives at the town of St. Mary's.

Immediately on my arrival at St. Marks, I communicated Mr. Bowles's design of taking that place [Fort San Marcos] to the Commandant, and two or three days after forwarded despatches to Governors White and Folch, with copies of the foregoing letter to Col. Hawkins, in order that they might be on their guard, and that the officers of the United States might not be accused of improper conduct, or duplicity. But the caution had no effect: Mr. Bowles remained unmolested, until he had in part regained his former influence with the Indians, and then besieged and took the Fort of St. Mark's [May 21, 1800], which was defended by about one hundred infantry, and more than one dozen pieces of good artillery.

At St. Mark's, I was treated by the Commandant Mr. Portal [Tomas Portell] and his lady, with politeness and hospitality. Madame Portal is an agreeable Spanish lady, and possesses a considerable share of vivacity, and good understanding. She speaks the English language tolerably well, and assisted an interpreter in a verbal communication which I received at that place relative to the money that was put on board a boat at New Madrid, for certain citizens of the United States, as already stated: the names of the persons for whom the money was intended, and for whom the receipts were given by the person who had the distribution of it, were likewise communicated. One of Mr. Panton's clerks was present when the communication was made.

Fort St. Marks, (frequently called Apalachy,) is situated on a point of land at the confluence of the Apalachy, and another stream of nearly the same size; they are too small to be called rivers. The Fort is built of hewn stone, and the work tolerably well executed; on the north side of the Fort, and adjoining the wall, is a deep wet ditch, which extends from one of the streams of water to the other.

The country round the Fort, is almost as level as the water in the Bay, and but little elevated above it: so that when the tides, which rise but between two and three feet, are aided by a brisk S.W. wind, it is overflown. This low tract of country is covered with high grass, and has every appearance of being a deep marsh; but this is not the case; the soil generally, does not appear to be more than two or three feet deep; and in many places much less, and is supported by an horizontal stratum of lime-stone of an indifferent quality. The Fort is built of this stone, and likewise an old tower, which stands about a league S.W. from the Fort. The foundation of the tower, is evidently on this extensive stratum of calcarious matter. The quarry from whence the stones were raised to build the tower is but a short distance from it.

When, or for what purpose, this tower was erected appears to be uncertain; but from its mouldering condition, and the total decay of the wood formerly connected with its walls, it is probable that it was built shortly after the Spaniards took possession of that country. On the top there appears to have been a light-house; but from the condition of the bay, being so shoal and full of oyster banks, as to render it useless to ordinary shipping, it appears to have been unnecessary. There is about half an acre of ground round the tower, raised by art so high, as not to be overflown, which was probably used as a garden.

The stratum of lime stone extends northerly a considerable distance, and is possibly connected with those extensive strata on the Chattahocha and Flint rivers, which may be considered the base of a large tract of sandy country. I frequently observed rocks of it, standing several feet above the poor sandy soil, (which appeared only adapted to the production of pitch pine and wire grass,) on the path between the lowest Indian village on the Chattahocha, and the boundary.

The principal source of the Apalachy, is from a single lime stone spring; many others are to be met with in that sandy country almost as remarkable for their magnitude; but they all issue from lime stone. The first lime stone observed on the boundary, east from the Mississippi, was between the Conecuh and Chattahocha rivers.

Some miles north of St. Marks, there is a tract of country, though not extensive, which is tolerably good, and here the Spaniards had a small settlement or colony; but it was conquered about sixty years ago, by an enterprising party from Charles Town, South Carolina; it is now totally abandoned, and scarcely a vestage of the settlement remains, except the ruins of a fort, and one or two pieces of old artillery, almost in a state of complete decomposition.

The latitude of St. Marks, by two good observations appears to be 30°8′29″ N. I had no opportunity owing to bad weather, of making a single observation for the longitude.

I expected to have been overtaken at St. Marks, by a vessel laden with a quantity of provision from New Orleans, which had been deposited at that place by our contractor. Supposing this vessel would pass through St. George's Sound, and if so, be liable to be captured by Lieut. Wooldridge, and his men, who had saved their boats. I mentioned the circumstance of this vessel, and provision to the Lieutenant, his officers, and Mr. Bowles, while with them on St. George's Island: and requested them to furnish her with a passport to follow me round Cape Florida, to St. Mary's if I should have left St. Marks before she arrived. They were likewise informed that no objection would be made to their taking such a supply of provision as their immediate necessities required, provided they gave the master a receipt for it. They appeared to be highly satisfied and returned me many thanks, and in the most positive terms, assured me that the vessel should not be detained if she fell in their hands; but forwarded immediately agreeably to my request, after supplying themselves with some necessary articles.

The vessel, with the provisions not arriving at St. Marks so soon as I expected, and the stormy season setting in, it was thought best to proceed without waiting longer for her. On the 16th of October, our commissary, Mr. Anderson, procured some fresh beef, which he had salted, and barreled up, and we made ready to set sail the day following.

My undertaking this voyage, was a matter of necessity, and not of choice, and could it have been avoided with advantage to my country, I certainly should not have taken upon myself so important a charge. Having on board the commissary Mr. Anderson with all his accounts, and vouchers, for the money expended since we left the city of Philadelphia in 1796: and all the papers, drafts, and astronomical observations, relative to the boundary, and some other important geographical positions, with the originals of all my correspondence, for more than three years, together with the apparatus, and baggage of the party, including the military escort: to which may be added, about twenty persons, of whom but five had never been at sea before, and of those five two only were sailors, and totally illiterate.

Thus circumstanced, I left St. Marks on the 18th of October, in a small light built schooner, of not more than 38 or 40 tons burden. The water being low, the vessel stuck some time on the oyster banks. At four o'clock P.M. we took our departure from the last stake in the channel, and at sun down were out of sight of land. About three o'clock in the morning of the 18th, we had a gale of wind from the east, attended with a dark mist, which obliged us to return into Apalachy Bay for shelter. The storm continued until the morning of the 20th, when we got up the yards [yardarms], and prepared to take advantage of the first fair wind.

On opening one of the barrels of beef, which had been salted up at St. Marks, it was found to be so much damaged, (with the exception of a few pounds,) as to be rendered useless: an examination was then made into the remainder, and it was unfortunately found to be in the same situation.

This discovery appeared to produce some uneasiness among the crew, several of whom were for returning to St. Marks for a fresh supply, but as we had a great sufficiency of bread and flour on board, the proposition met with such a decided negative, accompanied with a reprimand, that it prevented any complaints during the voyage, though we were frequently in disagreeable situations.

About eight o'clock in the evening of the 20th, we had a light breeze from the north, and got immediately under way, and put to sea.

21st. After midnight had a violent gale from the east; hove to under the double reefed foresail. About nine o'clock A.M., the wind abated, and shifted to the north. At ten o'clock P.M. from our soundings, it appear that we were in the great cove, or bay, between Cedar Keys, and Kayo Anclote, the latter are three small islands of St. Clement's point.

22nd. Still in the cover, and though out of sight of land until ten o'clock A.M. our soundings were commonly less than 10 feet, and in some places on an horizontal stratum of stone or rock, of a rough surface; on which was seen large patches of turtle grass, and sand, some of which is remarkably white, and large quantities of sponge. This horizontal stratum of stone, is probably connected with that already described at St. Marks. We arrived at the Keys in the evening, and came to an anchor between the northern one and the main land. In the night had a violent gale of wind, and were under the necessity of putting out our heaviest anchor.

23rd. Early in the morning got under way, and sailed to the anchor ground between the main land and southern Key, in an excellent harbor, where we intended laying until the weather became more settled. In the afternoon caught a number of fine fish.

24th. Very squally with dark mists.

25th. Weather the same as yesterday.

26th. Wind very violent from the S.E. until eight o'clock in the evening, when it shifted to the S. and brought on the heaviest fall of rain I had ever known. About midnight the wind changed, and came with such violence from the north that the vessel dragged both anchors a considerable distance.

27th. Wind moderated, about one o'clock P.M. when we immediately got under way and passed Hillsborough Bay (Spirito Santo Bay Tampe,) about seven o'clock P.M. This Bay I suspect is laid down in all our charts too far north, by at least fifteen minutes. Kayo Anclote opposite to St. Clement's Point, is laid down seventeen minutes too far north.

28th. The wind is fair. The sun was eclipsed whilst I was taking his meridional altitude.

29th. From a meridional altitude of Capella taken after midnight, we appeared to be opposite to Punta Larga, (Cape Roman,) and having a sufficient offing we steered for Cape Sable, the most southern promontory of east Florida, which was seen from the mast head at noon. At three o'clock in the afternoon came to an anchor on the west side of Kayho Anclote, or Sandy Key, which is a small island a very short distance south of the Cape. After coming to an anchor, myself and some of the crew, took our boat, and went to the island; where in a very few minutes, we shot about twelve plover. There are some bushes scattered over the island, but what particularly attracted my attention was the amazing piles, or stacks, of the prickley pear, (opuntia a species of the cactus,) the fruit was large and in high perfection: we eat very plentifully of it; but my people were not a little surprised in the next morning, on finding their urine appear as if it had been highly tinged with cochineal; no inconvenience resulting from it, the fruit was constantly used by the crew during our continuance among the keys or islands. Though this island is called Sandy Key, and has certainly the appearance of a body of sand; it is little more than a heap of broken and pulverized shells, which were found to effervesce freely with vitriolic acid, and little or no quartz was perceptible in the solution.

30th. Weighed anchor and sailed to Key Vaccas, or Cow Island, and moored in a small harbor among a cluster of little islands. Stormy all the afternoon. The soundings from Sandy Key, to Key Vaccas, were regular and generally less than nine feet, and on an horizontal stratum of stone, similar to that described between Cedar Keys and Kayo Anclote.

31st. Went on shore on Key Vaccas, where our people in a short time killed four deer, of that small species, common to some of those islands. They are less than our ordinary breed of goats.

November 1st. Examined a number of the small islands, they all appeared to be of lime-stone, or calcareous rocks, the tops of them were flat, and elevated but a few feet above the surface of the water, and covered with a thin stratum of earth. These rocks are evidently a congeries of petrefactions, in which may be traced a variety of plants, particularly the roots of the great palmetto, or cabbage tree, (coryphe or palmitto of Water). The mud in the harbor where we lay was of a fine white, and resembled lime, or whiting, and was found to effervesce with vitriolic acid; from which it is probable, that it is no more than shells, and other calcareous matter, levigated by the friction of the particles, produced by the constant motion of the water.

2d. Took some large turtle, and fine fish. Visited by Captain Burns of New Providence whose vessel lay at the east end of Key Vaccas. He was on a turtling and wrecking [salvage] voyage. Wind still from the east and squally.

3d. Killed some more small deer and salted them up. Calm the whole day.

4th. A light breeze in the afternoon, got under way, and proceeded about five miles along the north side of Key Vaccas. Soundings generally from seven to eight feet; the bottom horizontal rock with a rough surface.

5th. Got under weigh early in the morning, but the wind being ahead, come to an anchor under a small Key, a short distance from Duck Key. Soundings as before. On the small island there was some appearance of a clear field, manned the boat and went to examine it; but had proceeded but a short distance among the bushes, when I was compelled to return by the incredible number of musquetoes; on coming to the boat, I found the men had jumped into the water to avoid the attacks of those troublesome little animals.

This island is similar to those already described, but surrounded by a greater number of ragged rocks near the surface of the water.

6th. Got under way at eight o'clock A.M. and beat out into the channel between the Keys and reef, and came to an anchor in a good harbor at the east end of Viper Key.

7th. Made sail early in the morning, and came to an anchor at one o'clock P.M. in the harbor at the north east end of the Matacombe, where we found it necessary to take in wood and water. This island is noted for affording a greater quantity of good water than any other of the Keys; on which account it is much frequented by the turtlers [turtle hunters], and wreckers [salvage hunters]. The water is found in natural wells about four feet deep, which are no more than cavities in the rock, and not the effect of art as some have imagined. This island, like those already mentioned, may be considered as large flat calcareous rock, elevated but a few feet above the water, and covered with a stratum of earth. This is said to have been the last residence of the Coloosa Indians, the original inhabitants of East Florida: From whence they were gradually expelled by the Seminoles, or Wild Creeks. From Matacombe they were taken to the island of Cuba by the Spaniards, and incorporated with their slaves. But this measure does not appear to have been taken without provocation; These Indians were remarkable for their cruelty, which they exercised indiscriminately on all the unfortunate people, who were wrecked within their reach on that dangerous coast. The island of Matanza, (slaughter) which lies about one mile north east from the watering place, was so called from those Indians massacring about three hundred French, who had collected on it, after being wrecked on the reef.

On the north east side of Matacombe, there is a beautiful beach, which has the appearance of whitish sand, but on examination is found to be broken shells, coral, etc.

8th. Spent in taking in wood and water. In the afternoon the schooner Shark, late the property of Messrs. Panton, Laslie [Leslie] and Company, of Pensacola arrived; being a prize to Lieut. Wooldridge who captured her on her way to Apalachy. The schooner Shark was loaded with provisions, and as we had not meat, our commissary Mr. Anderson made application to the prize master for a barrel of pork; the prize master Mr. Barnet made no direct answer, but said he would see about it the next day.[55]

55. No mention of "General" Bowles is made here, and one can only conclude that he had returned to Florida with his staff, per the British plan, in order to once again stir things up among the Indians against the Spanish.

9th. Got under way at nine o'clock A.M. The schooner Shark did the same. It was previously agreed, that we should anchor together in the evening at Key Rodriguiz, where the prize master was to furnish us with a supply of meat, in return for what flour they had from us on St. George's island. We sailed rather the fastest, and so soon as we came to an anchor in the harbor, she crowded all her sail, and stood over the reef for New Providence. Thus were we requited for our favours. Soon we were joined by Captain Watkins, who commanded a privateer from New Providence. He behaved with politeness, and furnished me with about five pounds of excellent salt pork.

10th. Wind ahead, were not able to make any way. Our men caught a number of fine fish.

11th. Calm until about eleven o'clock A.M. when we had a light breeze and immediately got under way, proceeded to Key Large, and came to an anchor between the Key and Gulf Stream. At the same time a sloop that we were meeting, came to an anchor about two leagues from us.

12th. About two o'clock in the morning, I was called up to see the shooting of the stars, (as it is vulgarly termed,) the phenomenon was grand and awful, the whole heavens appeared as if illuminated with sky-rockets, flying in an infinity of direction, and I was in constant expectation of some of them falling on the vessel. They continued until put out by the light of the sun after day break. This phenomenon extended over a large portion of the West India islands, land was observed as far north as St. Mary's, where it appeared as brilliant as with us. During this singular appearance, the wind shifted from the south to the north, and the thermometer which had been at 86° for four days past, fell to 56°.

Many ingenious theories have been devised to account for those luminous and fiery meteors, but none of them is so satisfactory to my mind as the conjecture of that celebrated chemist M. Lavoisier, who supposes it probable that the terrestrial atmosphere consists of several volumes, or strata of gaz or elastic vapour of different kinds, and that the lightest and most difficult to mix with the lower atmosphere will be elevated above it, and form a separate stratum or volume, which he supposes to be inflammable, and that it is at the point of contact between those statra that the aurora borealis, and other fiery meteors are produced.

About eight o'clock in the morning, the sloop we saw the preceding evening passed by our stern, and upon being hailed answered, 'it is prize,' she was then ordered to come to, to which a person answered 'ay,' and at about 400 yards from us, hove to, and brought her boat, which was in tow along the side; but contrary to our expectation, it was immediately taken in, and the sloop with all her sails set, bore away. Orders were then given to get under way, and give chase, from an idea that it was an American vessel taken by the French, and if possible to retake her, several of her people not having the appearance of Americans or Englishmen. As soon as we got under sail, a gun discharged towards the sloop, but to which no attention was paid, but in about one hour we came within rifle shot, when one was discharged, and with such a direction that convinced the crew their safety depended upon coming to; which was immediately done, and we passed under her stern: The master was requested to come to anchor, and bring his papers on board of us. We anchored about rifle shot from the sloop, after which the request was repeated, but one of the persons on board the sloop observed, that the sea was rough, and they had but one oar and a paddle for the boat: upon which our commissary Mr. Anderson took the boat belonging to our vessel, and brought the master and his papers on board. The papers were satisfactory. The vessel and loading were lately Spanish property, and had been taken about fifteen days before by a New Providence privateer near the Havannah [Cuba], and sent on for Nassau; but got becalmed in the Gulf Stream, which carried her almost to Cape Carnaveral [Canaveral] when the wind served, the master then kept [to] the Florida coast until we met with him. He and his people had been seven days on allowance of one biscuit, and a pint of water each per day, and what fish they could take, which had to eat without salt. The master took breakfast with me, and when he was ready to return, I directed our commissary to furnish him with a barrel of biscuit, and some salt, upon which he observed, that he had 'never before been so fortunately chased and taken.' One half of his crew consisted of the Spaniards taken on board the vessel, and they all equally had done duty. Immediately after this fruitless adventure, we got under way, and the wind began to blow with considerable violence, which gradually increased until we found it necessary to come to anchor, and were very fortunate in making harbour near the mouth of Black Ceasar's Creek.

13th. The gale continued with violence. Took some fish.

14th. The wind continued very violent until the evening.

15th. The wind violent from the north, until one o'clock P.M. when it shifted in a few minutes, and came from the east; which was the only wind from which we were not protected by shoals, and which would in a short time have rendered our situation extremely uneasy. We got under way as possible; and beat out in order to fall into the northern channel of Black Ceasar's Creek; but having the wind and a strong current against us; we did not clear the shoal between the two channels until a few minutes before sunset, and then took the northern channel, which is very narrow at the entrance, not exceeding fifteen yards wide, but gradually widens to more than one hundred, and has between two and an half, and three fathoms water, except at the entrance where there is but seven or eight feet. We came to an anchor near the mouth of Black Ceasar's Creek, which is only the entrance into an extensive sound between the Keys and main land. The sides of the channel are almost perpendicular, like those at old Metacombe, and composed of a soft, whitish mud, which appears to wholly calcareous.

16th. Capt. Watkins beat up to us; he was the whole day making two leagues, in a vessel calculated to sail on a wind. He had with him the crew of the prize before mentioned: the vessel was wrecked by the violence of the wind the day we left her.

17th. The wind still continued very unfavourable; Took a considerable number of fine fish.

18th. The wind was more moderate, and we got under way early in the morning, and beat along Hawk Channel. In the afternoon were brought to by a New Providence privateer, commanded by Captain William Ball, who had been a short time from Ireland, and who treated us for some time with a degree of insolence far beyond anything I had ever before experienced. But after examining my instructions and commission, and viewing the signature of President Washington with all the attention and veneration that would have been paid to a holy relick, he became more moderate, and made us sufficient compensation for his insolence, by presenting us with a fine turtle, and after wishing us a pleasant passage, we parted.

About sunset there was an appearance of a storm, and we came to an anchor in a small, but excellent harbor, where we were defended by shoals from violence of the sea on every side: before midnight the storm came on.

19th. The storm continued the whole day.

20th. The storm still continued.

21st. Very strong gale from the N.E. Saw a ship early in the morning, (which had certainly missed her way,) nearly on the reef, and in very great danger, but she fortunately wore off.

22d. Got under way, and beat along the sound to the mouth of Fresh Water river, which is nearly opposite the southern part of Key Biscanio.

23[d]. Went on shore at the mouth of the river, filled our water casks, and gathered a large quantity of very fine limes: a party of our people likewise took their rifles, and went into the country, and were uncommonly fortunate in killing deer and turkies.

Fresh Water River is said to be no more than the outlet to a large lake, but a few leagues distance from the coast. At the mouth it is not more than five or six perches wide, and ten or twelve feet deep, and middling rapid. The sides are nearly perpendicular, and composed of calcareous stone or rock, similar to that described at Apalachy. The stratum of stone appeared to be very extensive and horizontal.

Key Biscanio is one of the last islands on the reef, and situated in lat. 25°37′ N.

The Florida reef, (as it is called,) appears to consist of a number of coral banks on the outer edge of an extensive stratum of calcareous stone, which extends from the main land, to the edge of the Gulf Stream: the general position of this stratum is nearly horizontal, and is possibly a continuation of that observed at Apalalchy. If this should be the case, it may be considered as the base of East Florida, and conform to the general law observed in the disposition of the strata of stone on our western waters.

On this stratum of stone, which serves as a helmet to the southern promontory of East Florida, and defends it from the violence of the Gulf Stream, is situated the whole of that cluster of innumerable islands and shoals, which have been so troublesome and dangerous to navigators.

These islands and shoals, may be viewed as protuberances, (standing on the surface of this extensive stratum,) gradually formed during a period of many centuries, by the constant accretion of calcareous matter. Many of those islands and shoals have evidently had their origin from coral banks, which not only like those of oysters, are known to increase, but to surpass them greatly in magnitude: and it is now reduced to a certainty, that a number of the islands in the South Sea are coral rocks covered with a stratum of earth. It is likewise well ascertained by naturists, that coral is not, as was formerly supposed a vegetable substance, but a vast collection of small animals which build up those rocky edifices from the bottom of the ocean!

The navigation between the Gulf Stream and Florida Keys, has at all times been considered as very difficult and dangerous, which it certainly is for those not acquainted with it; but with a competent knowledge of the Keys and reef, added to ordinary caution, I know of none more safe for coasting vessels, and others drawing not more than nine feet water. Such vessels are sailing from the northward into the Gulf of Mexico, and prefer the passage between the Gulf Stream and Florida coast, after entering the reef a few miles north of Key Biscanio, should be careful to give that Key a birth of about one and an half mile, on account of a shoal that makes out from it: it will likewise be necessary to observe, that opposite to the south end of the Key, there are but eleven feet water.

After entering the reef, it will be proper for a careful person to be kept aloft, who will be able for a considerable distance, (at least one mile,) if the weather should be fair, to discover the coral banks, rocks and shoals, which in some places are numerous, by which means the danger may easily be avoided. It will likewise be necessary on coming to an anchor, which must be done every night while on the reef, to look out for clear ground, otherwise a cable may be fretted off in a few hours by the coral rocks, or other protuberances [bulges], and the vessel go adrift.

As a knowledge of this navigation is of very great importance to the mercantile interest of the United States, it is a subject of regret that we have no charts in common use of the reef and Keys, (or islands,) upon a scale sufficiently large and accurate, to be useful. Mr. Gauld's survey of the Dry Tortugas and the Florida reef and Keys, easterly to Key Largo, made by the direction of the Board of Admiralty of Great Britain, may justly be considered as one of the most valuable works of the kind extant, but unfortunately it is little known. From Key Vaccas to Key Largo, I carefully compared Mr. Gauld's charts with the soundings, and perspective views of the Keys; and found an agreement which excited my surprise, and am induced to believe that not a single rock or shoal, so far as the work extends, has been omitted, and that not an error of three feet will be found in any of the soundings. If this work had been completed, it might be esteemed one of the most perfect and useful of the kind. The copy which I had the good fortune to obtain, (and without which it would have been very difficult for me, being not only a stranger to the coast, but no seaman, to have made my way with safety,) I deposited since my return in the office of the secretary of the navy.

Along the Florida Reef, and among the Keys, a great abundance and variety of fish may be taken: such as hog-fish, grunts, yellow tails, black, red, and gray snappers, mullets, bone-fish, amber-fish, margate-fish, barracoota, cavallos, pompui, groopers, king-fish, siber-fish, porgys, turbots, stingrays, black drum, Jew fish, with a prodigious variety of others, which in our situation we found excellent. Turtle also to be had in plenty; those we took were of three kinds: the loggerhead, hawk-bill and green; the two last are much the best. We likewise found a remarkable species of prawns, which live in great numbers in holes in the rocks: they frequently weigh two or three pounds apiece, and are improperly called lobsters; they want the large claws that lobsters have. Their meat is harder, and less delicate than that of the lobsters of the northern states.

Some of the Keys or Islands, were formerly very well timbered, but the most valuable kinds, such as lignum vitae, fustick and iron wood, have generally been cut off by the inhabitants of the Bahama Islands.

Key Biscanio is much frequented by the privateers, wreckers and turtlers from the Bahama Islands. At the south end there is an excellent harbor, and the shore so bold that a vessel not drawing more than ten feet water may be careened with safety. In that harbor we found several of those privateers, wreckers and turtlers, by whom we were politely treated, particularly by a Capt. Johnston, who furnished me with seven or eight pounds of salt pork.

Having filled our water casks, salted up some fish, and the wind serving, on the 25th, about noon, we got under way, and proceeded over the reef into the Gulf Stream. Shortly after we had entered the Stream, we saw a vessel bearing down upon us, but did not discover that she was a privateer until she attempted to bring us to by a shot: being determined to make the best use we could of the first fair, strong breeze we had had since our arrival at the Keys, we crowded all our sail, and the privateer did the same, but in two hours she gave up the chase.

In the evening we had a sudden and violent gale from the west, which laid the vessel almost on her beams before the sails could be handed: continued our course north, under the gib and double reefed mainsail.

26th. At break of day the wind shifted to the north, and continued with such violence that we had to heave to under a balanced main-sail. By a good observation our latitude 27°2′ N.

27th. The gale continued until about noon, when the wind shifted to the N.E. and became moderate. By a good observation our latitude 28°12′. From which it is evident we made our way good 1°10′ in twenty-four hours against a very strong gale, whilst laying to under the balanced main-sail, in which situation our drift would have been at least 1½′ south, the current must therefore have a set 1°46′ north during that time, or at the rate of nearly 5.1 English miles per hour! Such is the effect of this surprising stream. Being convinced from the last observation, that a N.W. course would carry us clear of the northern pitch of Cape Carnaveral shoal which is laid down in our charts in lat. 28°18′ N. that course was sailed on with a light breeze until four o'clock P.M. when we were becalmed. At eight o'clock P.M. a light breeze from the east, made sail to the west, in order to get clear of the Stream, which north of Cape Carnaveral sets north-easterly. At ten o'clock P.M. becalmed.

28th. At one o'clock A.M. a light breeze from the north: made sail to the west. Just after sun rise had soundings in eighteen fathoms on red shells: at this circumstance I was not a little surprised, as red shells are not met with much north of Cape Carnaveral, and as we had evidently been drifting seven or eight hours in the Stream, it was taken for granted that we were north of the Cape, eight or nine leagues. Continued west; kept the lead going, and gradually shoaled to ten fathoms, when land was seen from the mast head: the same course was continued until we had seven fathoms water, and a point of land to the north; bottom still red shells. Our latitude at that place by a good observation was 28°14′. Our course west, was continued until the observation was worked, when we had 10 fathoms and muddy bottom. It was then certain that we had passed over the southern part of the Cape shoal, and that the point of land to the north, was the Cape itself. From this it appears evident, that after leaving the Gulf Stream, we must have fallen into a strong counter current setting southerly, have made but two minutes northing in twenty-four hours, seven or eight of which were driving northerly with the Stream. A case of the same kind I find related by Mr. De Brahm. These currents on the coast baffle all calculation, and doubtless occasion, the loss of many vessels for which the masters are unjustly censured.

Becalmed from one o'clock P.M. until seven, when we had a brisk breeze from the S.W. made sail to the east for three hours, in order to clear the dangerous shoal off the Cape, and then stood north until midnight.

29th. From midnight until two o'clock A.M. stood N.W. to regain the coast; but the wind then shifted to the N.W. and became so violent that we hove to under the double reefed fore-sail. At daylight, to our mortification, we found ourselves in the Gulf Stream. At noon our latitude appeared to be 29°2′ N. In the afternoon the vessel shipped a sea which parted the lashing of one end of our boat, and nearly carried it off the deck, broke one of the stanchions, and injured the hand rails; handed the fore-sail, and laid to under the balanced main-sail.

30th. The gale still continued with great violence, and the appearance of the sea was alarming, and though our vessel laid to with east, and labored but very little, the main deck was constantly covered with water, and the seas broke over us with such rapidity for some hours that I was seriously apprehensive of foundering.

By a good observation found our latitude to be 30°10′ N. from which it appears that we made 1°8′ northing during the preceding twenty-four hours whilst laying to against a violent N.W. gale, in which, had it not been for the stream, our drift to the S.E. could not have been less than 1½ miles per hour.

About three o'clock in the afternoon the wind shifted to the north, and became less violent. At eight o'clock in the evening John Ransom the vessel's cook died, after about two weeks illness: he was a soldier in the revolutionary army of the United States, and continued in the service of his country, until discharged on account of his age at Loftus's Heights [renamed Fort Adams] on the Mississippi. From thence he descended the river to New Orleans, to procure if possible by his labour a scanty pittance for the remainder of his days under a despotism, after devoting the prime and vigour of his life in assisting to establish the freedom and independence of the United States! At New Orleans I took him into employ through charity, and to gratify his wishes to return and die in the land of liberty for which he had fought.

About nine o'clock P.M. the wind had fallen so much, that we were able to make sail to the west under the gib, and double reefed main-sail. This night we had ice upon the decks.

December 1st. Wind shifted to the N.E. but very light, continued our course west. At ten o'clock A.M. committed the body of John Ransom to the sea.

By a good observation our latitude was found to be 30°50′ N. Becalmed in the afternoon. Light breezes from various quarters during the evening, but generally from the southward, and we were not able to make good our course to the west.

2d. Soundings at two o'clock A.M. in eighteen fathoms, the bottom black and white sand, with some brown shells. The whole of this day we had gentle breeze from the S.W. At ten o'clock soundings in ten fathoms water, when land was seen from the mast head. By a good observation our latitude 31°13′.

Not having the latitude of St. Mary's in any publication, but from a small mutilated chart it appear to be about 31° N, it was therefore expected that our course would carry us to the north end of Cumberland Island, the wind not permitting us to lay more south. At four o'clock in the afternoon came to an anchor at the north end of St. Simon's Island in a good harbor. The satisfaction which the crew, and myself experienced on this occasion may be more easily conceived than expressed. We were now able to take a night's repose, free from those cares, and anxieties which must ever attend the reflecting mind in our past situation, exposed to the turbulence of the sea in a little vessel, having but two young illiterate sailors on board, along a dangerous coast with which we were all unacquainted, and experiencing three violent gales of wind, which we afterwards found had wrecked as many vessels, much better calculated to resist the fury of the winds, and billows than ours. So great was the dependence on observation, and so little on dead reckoning among the currents near the coast, that the log was never hove once during the passage.

My reason for being so minute in noting the results of the observations, and the direction of the winds, with the courses we steered after leaving Key Biscanio, was for the purpose of giving some idea of the velocity of the Gulf Stream, from a retrospect of which it will appear that while we were in it from Key Biscanio, to about latitude 30° it must have set northerly from 4 to 5½ miles per hour; but it may be necessary to observe, that it is not regular, it has frequently been found less and sometimes more.

The advantages to be gained by vessels taking the stream, which are sailing from the Gulf of Mexico northward, are too obvious to merit a single remark. From Key Biscanio, the course of the stream is a few degrees east of north, until it comes opposite to the shoal of Cape Carnaveral, from thence it forms a curve, still inkling more and more easterly to the shoals of Nantucket, and from thence, it is said, towards the Western Isles until it loses itself in the Atlantic.

In sailing south into the Gulf of Mexico, the stream should be avoided with as much care as it is sought for in the other case. The stream in its progress from the Florida Keys, forms three large eddies on the coast: the first south of Cape Carnaveral, the second between Capes Carnaveral and Hattaras, and the third between Cape Hattaras and Nantucket shoals. Such vessels as are sailing to the southward between the coast and stream, will frequently be benefitted by the eddy currents setting southerly. In taking this passage, it will be proper to give the shoals of both Cape Hattaras, and Cape Carnaveral good births; particularly the latter, which is said to extend eight leagues from the Cape, and so soon as they are cleared haul in a little for the land. In latitude 26°30′ N. the stream passes near the land, and sometimes touches it. The passage between the stream, and the coast was much used by small vessels, while the British nation held the Floridas.

The breadth of the Gulf Stream is somewhat uncertain, and has been differently estimated by different writers; but by taking a mean of the several determination, it will probably be near the truth: Between Key Biscanio and the Bahama Banks, it is about fifteen leagues broad; in latitude 28°30′ N. about 17. Opposite to Charles Town 20°. Off the Capes of Virginia 31 and, in latitude 40° about 38 leagues broad. As the stream proceeds northerly and eastwardly, it is naturally increased in width, and diminished in velocity: But its velocity from various causes is considerably accelerated or retarded, and as those effects are produced by causes operating in a different part of the ocean, the changes in velocity are not subject to any regular calculation.

Various theories have been devised to account for the phenomenon of the Gulf Stream. By one, the Gulf of Mexico is considered as a great whirl-pool, occasioned by the water being thrown into it between the western extremity of the island of Cuba, and Cape Catoch by the trade winds and tides, and thrown out by a rotary motion between East Florida and the island of Cuba, where it meets with the least resistance. By others it has been attributed to the water thrown into the Gulf of Mexico between the west end of Cuba and Cape Catoch, by the trade winds alone, and making its way out through the Gulf of Florida, where it is the least obstructed.

The latter theory scarcely merits a discussion, for it must be evident that though the winds are for the most part easterly within the torrid zone, yet whenever calms happen in the West Indies, and south along the coast which are not uncommon, the water must recede back to restore equilibrium, and not only cease to be pressed into the Gulf of Mexico, but rush out where it had before been pressed in, and an equilibrium take place between the Gulf and the Ocean, which is never the case.

The first theory appears to be correct in part, for it is impossible upon any principle of hydrostaticks, to account for the Gulf Stream without admitting a rotatory motion of the waters; but the centre of this rotatory motion is no more in the Gulf of Mexico, than the earth is in the centre of the solar system, and one is not more absurd than the other. I had an opportunity of examining the coast of the Gulf of Mexico from the outlet of Lake Pontchartrain, to Florida Point, and neither the currents, nor any other appearance would justify a supposition that the Gulf had any more similitude to a whirl-pool, than our lakes which supplied with water at one place, discharge it at another.

It will be evident upon a moment's reflection, that the vast body of water carried northerly and easterly by the Stream, must in some manner be returned southerly and westerly: if this be taken for granted, it follows of course, that the Atlantic Ocean, or a port of it, must have a rotatory motion about some centre within itself.

By admitting this circular motion in the water of the Atlantic, and though this motion be but small, it will nevertheless in a great degree be sufficient to account for the phenomenon of the Gulf Stream.

The water in its circular, or rotatory motion, is thrown upon the coast of America a little north of the equator, where from its centrifugal or projectile force, it becomes a little elevated, and still being carried along the coast northerly and easterly, on which the water continues from the same cause to be thrown, and at length meets with another body somewhat elevated, and upon the same principles carried westerly along the southern coast of the Island of Cuba, until at length this column of water so united, and thus set in motion, constantly contracted in width, and proportionably elevated above the true level of the sea, is brought as it were to a focus between the western extremity of Cuba and Cape Catoch, where it discharges itself into the Gulf of Mexico, which serves as a great reservoir, and contributes to the uniformity of the Gulf Stream. The water thus thrown into the Gulf of Mexico, issues out between East Florida, and the eastern part of the island of Cuba and the Bahama Banks, where the water of the Ocean is less elevated.

The quantity of water thrown into the Gulf of Mexico, is no doubt considerably increased and diminished by the different courses of winds and calms; but never so much diminished as to render the velocity of the Gulf Stream inconsiderable; which would certainly be the consequence if the cause depended immediately on the winds.

Several writers have remarked that the water of the Gulf Stream is five or six degrees, (and sometime more), of Farenheit's thermometer warmer than the adjoining water in the Atlantic, and between the Stream and our coast. This is certainly true, and thereby furnishes an easy method for seamen who are accommodated with those useful instruments to determine when they are in the Stream.

This difference of temperature arises from the water in the stream remaining a considerable time near the equator, and then flowing with rapidity into a colder climate, and though as it proceeds northward, it continued to lose its heat, it is nevertheless passing through water which still becomes colder, as it advance north: so that the relative difference continues nearly the same for a great distance. The difference which I generally found between the water in the stream, and the eddy water on the coast, was about seven degrees.

It has been supposed by some ingenious writers, that because, after leaving the Stream, and having soundings on our coast and a diminution in the heat of the water about the same time, it followed of course, that the water on soundings and banks is always colder than the adjoining. Though this may constantly be the case on our coast, it is probable the conclusion ought to be considered a particular, and not a general one. On our coast the Stream passes nearly along the great bank of soundings, it is therefore very natural to suppose, that soon after leaving the Stream, you will have soundings, and be in one of the large eddies on the coast whose waters being nearly stationary, and therefore colder than that moving with rapidity from the southward.

Again, it may be observed that the adjoining water in the Atlantic without the Stream is also colder, as well as that on soundings; but on the contrary fathomless. Hence the difference in this case, does not appear to depend upon the depth of the water, but upon the current setting rapidly from a warmer into a colder climate. From this a conclusion may very fairly be drawn, that the sudden changes found in the temperature of the water in the ocean, are more immediately the effects of currents, than of banks and soundings; but as those currents are general near coasts, and frequently occasioned by them, the thermometer may be considered a good monitor.

It has been mentioned by Dr. [Benjamin] Franklin, that the water of the Gulf Stream does not sparkle in the night. This, so far as my observations go, is incorrect. I saw little or no difference between that and the other water on the coast; but if there was any, that of the Gulf Stream was the most sparkling and luminous: It may however , be observed that the same water is very different, at different times in this respect.

The same ingenious writer and philosopher, likewise observes, that the Gulf weed is a sign of being in the Stream. This is in part true, but by no means to be considered a general rule, because the water on the borders of the Stream, is constantly mixing with the adjoining water, and leaving some of the weed behind, which consequently falls into the eddy currents, and is carried off many leagues: we met with it on soundings, in the eddy current, setting southerly. These remarks cannot effect the character of Dr. Franklin, either as a writer or philosopher: his character is formed of materials which will elude the destroying hand of time itself, and will be revered so long as liberty and science command the affections and esteem of mankind. I merely think the Doctor was mistaken, and conceive it my duty to state facts.

December 3[d]. Sent our boat to the island to procure some provision, and obtain information relative to the in-land passage to St. Mary's. By the return of the boat, we were informed that there was a good, and safe in-land passage to St. Mary's; but that we should have to employ a pilot. In the afternoon, the wind serving, we sailed into a large opening which we supposed led into the Sound.

4th. Before sunrise despatched the commissary Mr. Anderson to the village on St. Simon's to employ a pilot. A severe storm, with cold rain all the afternoon, and night.

5th. Clear all the afternoon with a strong gale from the S.W. In the evening the commissary returned with a pilot.

6th. The pilot removed the vessel into a harbor much less secure than where he found it. A heavy storm all night.

7th. Got under way, but the pilot soon ran the vessel aground between two good channels. At low water we had but eight inches at our stern, and exposed on one side to the sea. Finding the pilot extremely stupid, and unacquainted with the channels, as soon as the tide rose, and vessel was afloat, we moored into deeper water and came to an anchor. My own people then went, sounded the bar, and staked out the channel.

8th. Wind fair, weighed anchor, crossed the bar, and entered a branch of Alatamaha, which is connected with, and forms a part of the in-land communication between Savannah and St. Mary's.

9th. Arrived at the village of St. Simon's (properly Frederica,) about eight o'clock in the forenoon, and engaged another pilot. After engaging the pilot, I spend a few hours in examining the ruins of that once handsome and flourishing town; which during the whole or part of Gen. [James Edward] Oglethorpe's administration was the seat of the colonial government [after 1733] of Georgia.

The town of Frederica, as nearly as I could judge, appeared to have been regularly laid out, that is, the streets at right angles to each other, and the whole surrounded by a wall of earth, except that part lying immediately on the water, which was defended by a small battery of tabby work, (as it is called in the country,) which is a composition of broken oyster shells and lime. The walls of the principal houses in the town were also of the same composition. The appearance is similar to rough-cast; and some of the walls seemed as solid as though they had been cut out of rock.

The town was destroyed during our revolutionary war with Great Britain; and it is probable that it will not be rebuilt for many years to come; at present the population does not exceed twenty families.

The seat of Gen. Oglethorpe, the Governor, and protector of the colony in its infancy, was on St. Simon's Island, and not far from the town Frederica. Although the character and services of Gen. Oglethorpe, does not appear to be sufficiently appreciated, the monuments of his industry will long continue to do him that justice which his country denied him. While his time was employed either in defeating the Spaniards, (by whom his little colony was several times invaded,) erecting forts, joining the meandering waters in the low flat county by canals, to render the communications more expeditious, and certain, and otherwise increasing the value and consequence of his government, he was assailed both at home and abroad, by a host of complainers, and fault-finders, who having no intrinsic worth of their own, felt a pleasure in endeavouring to lessen that of others: And as it is much easier to find fault than to plan and execute, the illiberal, idle, and envious will ever be found attacking distinguished worth, industry, and talents, and too often be attended to, and patronized by those whose stations in government ought to render them inaccessible to the tongue of detraction, slander, and obloquy.

At half an hour after twelve o'clock, the tide serving, we left Frederica and arrived at St. Mary's at half past six in the evening. This passage was one of the quickest ever known. At St. Mary's we had the satisfaction of meeting with our companions, who came through the wilderness from the mouth of Flint River by land: their journey was tedious and disagreeable, on account of the autumnal rains then prevailing.

CHAPTER X.

Encamp at Point Peter and set up the instruments—Provisions scarce—Observations on East Florida—Provisions from New Orleans captured—Proceed higher up the St. Mary's, and encamp near Okefonoke [Okefenokee] swamp—Alligators, some particulars respecting them—Astronomical observations completed—The river St. Mary's described and the proper positions for military works on the Mississippi, etc., pointed out—Strictures on those already erected and on the state of military science in the United States—Botanical list—Conclusion.

When I arrived at the town of St. Mary's, the effects of Mr. Bowles return to the Creek nation was not known, in consequence of which a messenger was immediately despatched to our agent Col. Hawkins, to obtain some information on that hand, before we ventured to proceed into the country to determine the source of the St. Mary's river.

Finding that I could not obtain quarters in the town for myself and people, free of expense to the public, I moved on the 12th to Point Peter, and encamped in a forest of live oak, where a number of people were engaged in cutting timber for the United States: the official wood served us for firing. This system I pursued from the time I left Pittsburg in the year 1796, until my return to Philadelphia in the year 1800, and whatever attention, and shelter the men might require, I occupied no quarters myself at the expense of the public.

After we had encamped, the instruments were unpacked, and set up, and a course of observations, and some mathematical operations begun, which will be found in the Appendix.

We found provisions in that part of Georgia very high and scarce; and many of the inhabitants were importing corn, and other necessaries of life for their own consumption: from this circumstance we kept but a small supply by us, being in daily expectation of receiving that load from New Orleans, about which I had spoken to Lieut. Wooldridge and Mr. Bowles, when at St. George's island, and who had pledged their honour not to detain it, if it fell in their hands.

I shall now proceed to make a few observations relative to East Florida.

East Florida is but little better than a wilderness, the soil is not superior to that of West Florida, and none of its navigable waters rising in the United States, it does not appear equally interesting. It is nevertheless of immense importance to the United States, being from its peculiar situation, well calculated to give security to the commerce between the Atlantic and western states, and may be considered one of the main keys to the trade of the Gulf of Mexico. On the west side, it affords to remarkably fine harbours: one is known by the name of Hillsborough Bay, (Bay Tompa [Tampa Bay], or Spirito Santo.) The latitude is stated to 27°36′ N. and the longitude 83° west from Greenwich. It is very spacious, and will admit any vessel over the bar not drawing more than twenty-four feet water.

The first Englishman who explored, and gave an account of the bay was a Capt. Braddock, who commanded a privateer from Virginia, and cruised on the west coast of East Florida, in the years 1744 and 1745: his survey is yet considered as good as any extant.

The other harbor is called by the Spaniards Boca Grande, and by the English Charlotte Harbour, and stated to lay in latitude 26°43′ N. and 82°30′ west longitude from Greenwich. It has fifteen feet water on the bar, and good anchorage within. Exclusive of those harbours, there are several other well calculated for coasting vessels, that draw not more than seven feet water; but their situations are so badly determined, that an enumeration of them would be unnecessary.

The Florida Keys and reef, likewise furnish a great number of harbours proper for coasting vessels, and advantageous stations for cruizers; particularly that of Key Biscanio, situated at the northern entrance of the reef, and capable of commanding the whole coasting trade which should take that passage. This being the entrance of the reef, and the most proper place to depart from in sailing northerly, the whole coast, and perhaps on the continent for a light house.

But instead of any advantage being derived, either in the United States or his Catholic Majesty, from those favourable situations, they serve as dens and hiding places for the privateers and pickaroons [pirates] of the Bahama islands, by which the trade of both nations has suffered immensely in spoliations: and extraordinary as it may appear, it is no less true, that nearly the whole coast of East Florida, so far as maritime possession give a right, is under the dominion of the Bahama islands. The coast and islands being uninhabited even by a single solitary settler from Apalachy, almost round to St. Augustine! from which the inhabitants of the Bahama islands cut and carry off, without interruption, as much of the valuable ship timber as they find necessary or convenient.

On the east side of the coast south of St. Augustine, there are a number of small harbours, proper for coasting vessels; but their position are too badly determined to entitle them to attention.

We have not at this time, one chart of the coast of East Florida, except Mr. Gauld's survey of a part of the keys and reef, entitled to any confidence. The making a survey of the eastern side of it, was submitted by the British government, while his Britannic Majesty was in possession of that country, to M. de Brohm, and the west side to Mr. Gauld; but the labours of those gentlemen have never been communicated to the public! An accurate knowledge of the dangerous shoal off Cape Carnaveral, is of great consequence to the commercial interest of the United States. It frequently happens that those places, which from the want of a competent knowledge of them are avoided, when critically examined will be found to afford places of safety, and good harbours, for such vessels as are driven upon them by bad weather. Such was the case with Dry Tortugas until examined by Mr. Gould.

The discovery of East Florida is generally attributed to Juan Ponce de Leon in 1512; but it is probable the eastern coast was discovered about fifteen years before that time by Sebastian Cabot.

After the coast of East Florida had been discovered by Juan Ponce de Leon the country was visited by a number of adventurers; but the first patent was obtained by Francis de Geray, who did not live to take possession of the province. Francis de Geray was succeeded by Luke V. de Allyon, who visited Florida about the year 1524, and was succeeded by Pamphilo de Narvaez in 1528, or 1529, who died on the coast, and was succeeded by that celebrated adventurer Ferdinando de Soto; who traversed both the Floridas, and part of our western country from the year 1529, to 1542, and died at the forks of the Red River, or some writers state on the Mississippi.

The first permanent settlement in East Florida, was attempted by some French protestants in the year 1562, to secure to themselves a retreat from religious persecution. But as soon as the king of Spain received an account of the commencement of this infant settlement, he despatched Don Pedro Malendez [Menéndez] de Aviles, into East Florida, with a considerable force to destroy it, which he effected in a most cruel and barbarous manner, in the year 1565, and established a colony at St. Augustine.

For this service, it appears that Malendez obtained a grant for all Florida, which grant included the whole coast on the Gulf of Mexico, and as far north and east as Newfoundland; to which was added a number of privileges, for which he was to perform some signal services: one was to make a chart of the coast of Florida for the use of the Spanish navigators who visited those seas, but this service was never performed. Neither does it appear, that any measures were taken for that purpose until about 1718,when Don Gonzales Carrenza, the principal pilot of the Spanish flota [fleet] undertook it, but his observations remained in manuscript, and were little known, until published in London in the year 1740: they are however very imperfect.

In 1586 St. Augustine the capital of the province was taken and pillaged by sir Francis Drake, and in 1665 it was again taken and plundered by Capt. Davis, who headed and commanded a body of Buccaneers. In 1702 an expedition was carried on against it by Col. Moore, Governor of Carolina; his force consisted of five hundred English troops, and seven hundred Indians, with whom he besieged the city for three months without success, and then retired. Except [for] those incidents, the history of East Florida from the settling the colony in 1565, is little more than a succession of Governors, until Gen. Oglethorpe took possession of Georgia, which circumstance excited considerable jealousy at the court of Madrid, and a large force was sent against him, which he not only defeated, but after various encounters carried his conquests to the gates of St. Augustine, and laid siege to that city in 1740; but being badly supplied with almost every article necessary to give success to such an undertaking, he was obliged to relinquish his design.

By the peace of 1763, the Floridas were ceded to his Britannic Majesty George the third; but who in consequence of the ill advised war he made upon his American colonies, now the United States, and which involved France, Holland and Spain in the contest, was reduced to the necessity in 1783 of acknowledging the colonies independent states; and restoring the Floridas to his Catholic Majesty [1784], who yet retains them.

Notwithstanding the early discovery of East Florida, the interior of it yet remains but little known, and uninhabited, except by the wandering Creeks or Seminoles. It cannot therefore be expected as the Indian tribe is now of little importance, that his Catholic Majesty can draw any advantage from the province, to compensate for the expense he is at in supporting the government at St. Augustine.

On the 10th of January 1800, I received a letter from Mr. Panton, informing me, 'that when our provision arrived at Pensacola from New Orleans, it was discovered that the vessel was in bad condition, and the master unacquainted with the coast, he thought it prudent to have the loading taken out, discharge the vessel and hands, and forward the provision after us in his own vessel the Shark, which was unfortunately captured somewhere on the coast; but by whom he knew not.' This was the same vessel loaded with provision, that joined us at Matacombe as already related, and from which we were not able to obtain one single pound of meat, though it is now certain the provision was our own, and captured contrary to an express stipulation between Lieut. Wooldridge, Mr. Bowles and myself.

Upon receipt of the letter above mentioned, orders were immediately given to the contractor, to lay in a quantity of provision sufficient to serve the whole party, (including the military,) while ascending the St. Mary's river, and employed in ascertaining its source and geographical position.

On the same day that I received the letter from Mr. Panton, I also received one from our agent Col. Hawkins, in answer to mine from St. Mary's, by which it appeared, that he had not then understood how Mr. Bowles was received by the Creeks, and therefore prudently declined giving any opinion respecting the further prosecution of our business.

On the 19th the observations at Point Peter were closed, and the instruments taken down and packed up.

The 23d we left the town of St. Mary's and proceeded up the river as far as it was navigable for the United States schooner, and then made use of canoes until an end was put to our navigation on the 6th of February by drift wood, logs and other impediments.

Part of our journey up the river, after leaving the schooner and taking to canoes was extremely disagreeable, owing to bad weather, cold rains, and the wet marshy ground on which we encamped.

February 7th, we began our observatory, and sent a party to examine whether there was any communication between the river and Okefonoke Swamp, which after our arrival at St. Mary's to our surprise, we found doubtful. The same day a number of canoes were sent down to the vessel to bring up some of our instruments and other articles, we were under the necessity of leaving behind.

On the 12th the instruments and other articles arrived, and a course of observations was began as soon as the weather permitted. In the evening the party that was sent to explore the source of the river, or its communication with the Okefonoke Swamp returned; but without making any satisfactory discovery, and the day following another party was despatched on the same business.

This being the season that Alligators, or American Crocodiles, were beginning to crawl out of the mud and bask in the sun, it was a favourable time to take them, both on account of their torpid state, and to examine the truth of the report of their swallowing pine knots in the fall of the year to serve them, (on account of their difficult digestion,) during the term of their torpor, which is probably about three months. For this purpose two Alligators of about eight or nine feet in length were taken and opened, and in the stomach of each was found several pine and other knots, pieces of bark, and in one of them some charcoal; but exclusive of such indigestible matter, the stomachs of both were empty. So far the report appears to be founded in fact: but whether these substances were swallowed on account of their tedious digestion, and therefore proper during the time those animals lay in the mud, or to prevent a collapse of the coats of the stomach, or by accident owning to their voracious manner of devouring their food, is difficult to determine.

The Alligator has been so often, and so well described, and those descriptions so well known, that other attempts have become unnecessary. It may nevertheless be proper to remark, that so far as the human species are concerned, the Alligators appear much less dangerous than has generally been supposed, particularly by those unacquainted with them. And I do not recollect meeting with but one well authenticated fact of any of the human species being injured by them in that country, (where they are very numerous,) and that was a negro near New Orleans, who while standing in the water sawing a piece of timber, had one of his legs dangerously wounded by one of them. My opinion on this subject is founded on my own experience. I have frequently been a witness to Indians, including men, women and children, bathing in rivers and ponds, where those animals are extremely numerous, without any apparent dread or caution: the same practice is also pursued by myself and people without caution, and without injury.

Some of the Alligators we killed were very fat, and would doubtless have yielded a considerable quantity of oil, which is probably almost the only use that will ever be made of them; however their tails are frequently eaten by the Indians and negroes, and Mr. Bowles informed me that he thought them one of the greatest delicacies.

The Alligators appear to abound plentiful in musk, the smell of which is sometimes perceptible to a considerable distance, when they are wounded or killed; but whether the musk is contained in receptacle for that purpose and secreted by a particular gland or glands, or generally diffused through the system appears somewhat uncertain: and I confess their appearance was so disagreeable and offensive to me, that I felt no inclination to undertake the dissection of one of them.

The second party which had been sent to ascertain the connection, (if any,) between the river St. Mary's and the Okefonoke Swamp returned on the 17th, having discovered the communication, and the day following a traverse was begun, to connect the observatory with that part of the Swamp from whence the water issued, in order to determine its true geographical position. For the whole operation, with the observations, see the Appendix.

On the 25th the observations were closed, and a large mound of earth thrown up at the observatory.

On the 26th Capt. Minor his Catholic Majesty's commissioner and myself, with a party of labourers went to the Swamp, and the day following had a mound of earth thrown up on the west side of the main outlet, and as near to the edge of the Swamp as we could advance on account of the water. The next day we proceeded down to the vessel, and arrived at the town of St. Mary's on the 3d of March.

The astronomical part of the boundary between the United States and his Catholic Majesty, being now completed, it only remained to make out the report, with the maps or charts of the line. As a proper place for doing this business, we agreed to go and encamp on the south end of Cumberland island, where fire-wood could be had without any expense to the public, and where we could be more retired and less interrupted by company. Agreeably to this plan, we left the town of St. Marys' on the 6th day of March and encamped. The instruments were likewise unpacked and set up, that no opportunity should be lost for determining with precision, every important point on the coast, where our residence for a few days became necessary. The observations with the results are the last in the Appendix.

I shall now proceed to make some observations relative to the St. Mary's river, and the proper positions for military works within the United States on the Mississippi, and along the boundary from that river to the Atlantic Ocean.

The main branch of the river St. Mary's, which is part of the southern boundary of the United States, has its source in the Okefonoke Swamp, from whence the water issues along several small marshes or drains, which soon united in one. The Swamp which is also said to be the source of the river St. John's, which falls into the Gulf of Mexico, is certainly very large, though much less than has generally been supposed: it is watered by a vast number of small streams and drains, which generally rise within its vicinity. Exclusive of two rivers, it has also been the source of a number of ridiculous and fabulous stories.

The river St. Mary's is at all times navigable for topsail vessels as high as Trader's Hill, and would be from thence up almost to the Swamp for boats and canoes, when the water is moderately high were it not for logs, drift wood, and rafts [large floating masses of driftwood] , which in many places extend across the stream. A large branch of the river comes in from the west, above the place where we encamped and built our observatory, which is marked in the map. The river is extremely crooked, and a large proportion of its banks are annually inundated. The upland is generally of an inferior quality, producing little besides wire-grass, pitch-pine (pirus) and broom pine, (pinus palustris). I shall now proceed to note the positions for military works.

There are several places on the Mississippi between the mouth of the Ohio and southern boundary of the United States, that would answer very well for military establishments; but the best appear to be at Chickasaw Bluffs, Walnut Hills, and Loftus's Heights. As one of the three Bluffs above the Chickasaw Bluffs, but I cannot recollect which, a fort might be advantageously erected. Fort Prudhome (or Prud'homme,) was built upon the middle one.

It will be difficult to erect works on any part of the Mississippi, below the mouth of the Ohio, that will prevent the descent of troops. The rapidity of the water, and width of the river, will enable a boat with some exertion to pass any of the forts with but little, if any damage, and there is no place where a cross fire could be brought to bear with much advantage: but the ascent of boats and gallies is so slow, that a few pieces of artillery well directed and served, would stop the progress of any vessel employed on the river.

On the Pearl or Half-Way River, a very short distance above the boundary is a commanding eminence, where a fort might be erected that would easily prevent the ascent of such boats and periaguas [pirogues], as would be proper for that navigation.

My knowledge of the Pascagola is too limited to justify an opinion, but from its distance both from the Pearl and Mobile rivers, and direct communication with the Gulf of Mexico, added to its magnitude, I should suppose it worthy of as much, if not more attention than the Pearl river.

The Mobile, Tombeckby and Alabama Rivers, are at this time of much more importance to the United States than all the other waters between the Mississippi river and the Atlantic ocean: being the only rivers which are navigable for square rigged vessels from the Gulf of Mexico, into that part of the United States, lying on the north boundary of West Florida. But exclusive of this consideration, there is another, which arises from the lands on those rivers being already partially settled, and at this time the most vulnerable part of the Union.

The position of Fort Stoddard, on Ward's Bluff between the confluence of the Tombeckby and Alabama rivers, is a very proper one, but the works are neither sufficiently extensive nor strong, to oppose an enemy possessed of artillery, and so long as his Catholic Majesty holds West Florida, so long will it be necessary for the United States to be formidable in that quarter.

Any works on the Coenecuh river will be unnecessary for some time to come, there being no inhabitants on it to protect, nor a sufficient number of Indians residing on its water, to make that trade worth attending to. About one mile and an half above the boundary on the east side of the river, there is a place where a trader formerly resided, that would answer tolerably well for a small military establishment.

At the confluence of the Chattahocha and Flint rivers, the ground is swampy, and annually inundated, and therefore unfit for military works: but there are some Bluffs on the east side of the Chattahocha, which begin about one mile and three quarters above the mouth of the Flint river, where works might be advantageously erected.

On the St. Mary's river we have two military establishments, one at Colerain, and the other at the mouth of the river on Point Peter. Neither of them ever have, or will be of any advantage, either in protecting our trade, or adding to the security of our citizens: they possess neither advantage in situation, merit in design, nor strength in the execution. If a fort up the river were at any time necessary, the proper situation would have been at Trader's Hill.

In fixing upon Colerain, the maxim, that the interest of any individual is to yield to that of the community, has evidently been reversed, and the interest of the nation has yielded to that of the individual; the same may in part be said of the position at Point Peter.

The plan of the fort at Point Peter, and the execution of the work are not equally defective, but the very end, or design of a fortification appears to have been overlooked; being even deficient as a demi-lune [a crescent-shaped outer-work], to which it bears some faint resemblance, and covered by no other work. The embrasures [gun ports] are so injudiciously opened, that a vessel may come to an anchor opposite to one part of the work, and have sufficient room to ride to her cable, and not a single gun can be brought to bear upon her.

It may be objected, that the fortifications, (if they merit that term,) of the United States were merely intended to answer temporary purposes, and that there was no necessity of attending to that strength and accuracy in their construction, which would be proper in permanent works: this may be true, so far as it respects defence against Indians, for which purpose a block-house, surrounded with palisades is sufficient. But the case is widely different with scientific warlike nations and the causes which rendered temporary works necessary against them, may so frequently recur, that permanent ones would in few years be found the least expensive, and accord more nearly with the principles of national economy. On this subject one general's, though disagreeable, remarks will be sufficient: perhaps no civilized nation in the world, is as deficient in the knowledge of this important part of the art of war, as the United States.

An attempt was made during the administration of President Washington to form a military school [West Point Academy, chartered 1794], in which the scientific principles of fortification, and projectiles were to be taught; but the professorship, or command was given to a person ignorant of the common properties of a right line triangle, and the institution naturally fell into contempt. In Europe great care is taken to select the best informed scientific characters, to fill the professorships in their military schools, not merely for the purpose of instructing such as are intended for officers, but to make useful citizens, and subjects; for it is certainly a truth, as well established as any mathematical deduction, that the arts, sciences, and literature are the pillars of civilization, and open the way to ease, plenty, and comfort, and without which man must in a short space of time, occupy his primitive state of barbarism, ignorance, distress, and perpetual want; and it is no argument against his general position to say, that those acquirements are useless, because a portion of mankind in all civilized countries, are rich, and comfortable without having any knowledge of them, for such persons, notwithstanding their deficiency in those branches of knowledge, nevertheless enjoy in common their share of the benefits. Being insensibly led into this digression, I shall now resume the original subject.

The situation selected by the very judicious Gen. Oglethorpe, on the south end of Cumberland Island, where he erected Fort William, appears to me the most eligible, and better calculated for a permanent work to give security to the harbor, and sound, than any other position about St. Mary's.

It will probably be expected by some readers, and gratifying to others, to have a list of the indigenous plants, shrubby, and herbaceous, to be met with on the Mississippi, along the boundary, and in the Floridas; but being an indifferent botanist, I am constrained to be very limited on that subject, and shall note but few productions, which did not attract my attention either for their use, quantity, beauty or singularity. To the popular names, those used by the botanists are added, in the manner practiced by the ingenious learned, and truly patriotic author [Thomas Jefferson] of the Notes on Virginia. The botanical name constantly follows the popular one, and is included with a parenthesis.

At the mouth of the Ohio, and down the Mississippi swamp, one of the prevailing species of timber is cotton-wood, (populous deltoids of Marshall,) it bears a very striking resemblance to the Lombardy poplar, is equally quick in growth, open, soft and porous. Black willow, (salix nigra,) black-ash, (fraximus nigra,) sugar maple, (acer saccharinim,) but not in great abundance, and becomes more scarce as you descend the river. I do not recollect seeing but one tree of it south of the boundary. Water maple, (acer negundo,) pecan, (juglans illinoisaensis,) this is met with as high as the Wabash, where it is not scarce, but becomes more abundant from thence down to the Gulf of Mexico. Papaw, (annona triloba,) I have eaten of the fruit in great perfection as early as the 17th of July, in the Mississippi Territory. Button wood, or sycamore, (platanus occidentalis,) hickory, (juglans hickory of three species.) The cypress, (cupressus disticha,) begins to make is [its] appearance about the Arkansas, and becomes very abundant a little further south, and appears to be inexhaustible before you reach the 31st degree of north latitude. It occupies many parts of the swamp almost to the exclusion of any other timber. The cypress is very useful wood, and generally used in the country for covering, flooring, and finishing buildings. It grows in swamps, marshes and ponds, but not on dry high land. The stem, or body of the tree, generally rises from the apex of a large conical base, above which the workmen have frequently to erect scaffolds before they fall the tree. From the roots of the tree a number of conical excrescences grow up, which are called cypress knees; some of them are eight or ten feet high, and being hollow are used for bee-hives, and other purposes. The long moss, (tillandsia usneoides,) [Spanish moss] makes its appearance on the Mississippi nearly in the same latitude with the cypress, and almost covers some of the trees to which it is appended before you reach the Walnut hills. This is a very useful article, and answers almost as well for beds, and mattrasses, as craped [cropped] horse hair: it is nearly as elastic, and almost as incorruptible. Although this moss, is not like the mistletoe immediately connected with the trees on which it hangs, it will not live long on a dead tree. Sweet bay, (laurus bordonia,) magnolia grandiflora: this most splendid, and beautiful tree, I do not recollect seeing above the Walnut-hills; but have not doubt of its growing much further north. It is common through all the rich lands of Natchez, and east to the Atlantic. The foregoing appear to be confined to very wet, or very rich land, and will be met with in all such places along the boundary, and through the Floridas, with the exception of the pecan, sugar maple, and one or two others. The catalpa, (bignonia catalpa,) is not common; but appeared the most abundant on the banks of the Coenecuh. The nysisa aquatic, is common on the Chattahocha below the boundary. Exclusive of those plants, which are generally confined to low, or very rich grounds, the following will be met with in various parts of the country. Sassafras, (laurus sassafras,) which grows to a large size about the Natchez. Sweet gum, (liguid-amber,) common swamp gum, (syssa integriifolia,) holly, (ilex opaca,) in great abundance in some parts of the Mississippi Territory, and frequently becomes a large tree. Parsimon, (dyospyros virginiana,) very common. Locust, (robvinia pseud-acacia,) honey locust, (gleditsia triancanthos,) black walnut (juglans nigra,) elm, (ulmus americana,) dog wood, (cornus florida,) wild plum, (prunus Chickasaw,) tulip tree, (liriodendron tulipifera.) this is improperly called "poplar" in the middle states. White oak, (quercus alba,) black oak, (quercus nigra,) swamp oak, (quercus aquatic,) chestnut oak, (quercus prunus,) with several other species, or varieties. Live oak (quercus virens,) this very useful timber is much confined to the coast, and a short distance from it. I do not recollect seeing it in any considerable quantity in West Florida, as far north as the boundary. Red cedar, (juniperus virginiana,) this is likewise much confined to the coast, and in some places very abundant. Pine, (pinus,) broom pine, (pinus palustris,) with several other species, or varieties; the quantity inexhaustible. Buck eye, (aesculus pavia,) this is sometimes confounded with the buck eye (aesculus flava,) of the Ohio; they are not the same. Wild cherry, (prunus virginiana,) great palmetto, or cabbage tree, (coryphe or pamitto of Walter,) cassina yapon, (ilex vomitoria,) the black-drink used by the Creeks, and some other Indian nations at their councils, and public meetings, is made by an infusion of the leaves of this shrub. Myrica inodora, (of Bartram,) from the berries of this shrub, and green wax used in making candles is collected: these two last are confined to the coast. Beech, (fagus ferruginea,) chestnut, (fagus Americana,) chincopin, (fagus pumila,) some of them are sufficiently large in the Mississippi Territory to be split into rails. Spicewood, (laurus benzoin,) Bermudian mulberry, (callicarpa Americana,) cane, (arundo gigantean of Walter,) extends through all parts of the Mississippi swamp, and occupies equally the high, as well as low land, from the Walnut-hills down the river to Point Coupee, and easterly from fifteen to twenty miles or more. The whole of that high, rich, hilly and broken tract of country, except where the farms are opened, may be considered one solid cane-brake, and is almost impenetrable; but will probably be destroyed in a few years by cattle, hogs, and fires. Its general height is from twenty to thirty-six feet, though I have met with it on the tops of several hills forty-two feet high. The small cane, or reed, (arundo tecta of Walter,) begins to make its appearance on the boundary about twenty miles east of the Mississippi river, and with the arundo gigantean, or large cane, will be found on all the creeks and river bottoms, through to the Atlantic. The China root, (smilax China,) and passion flower, (passiflora incarnate,) are abundant in the rich grounds. The sensitive briar, (mimosa instia,) this beautiful and singular plant, is common to the poor, sandy land. Several species of that beautiful plant, the saracinia, are frequently met with in the margins of swamps, and low grounds; and three or four handsome species of the water dock, (nymphea,) poke, (phytolacca delcandra,) sumack, (rhus,) several species. Along the water courses, and in the swamps where the land is good, several species of well tasted grapes are found in great plenty. Many of the trees in the low grounds are loaded with a variety of vines, the most conspicuous of which are the creeper, or trumpet flower, (begnonia radicans,) and common poison vine, (rhus radicans,) mistletoe, (viscum,) is in great abundance, and will be found attached to almost all kinds of trees. I have frequently observed it on peach and plum trees. In the middle states it is generally found on the common swamp gum, (nysisa integrifolia).

To those who want more information on this useful and important subject, I would recommend the perusal of Bartram's Travels, a work which contains much valuable botanical knowledge; but from some circumstance, unconnected with its real merit and design, has not met with that attention from the public, to which it is just entitled.

Having now finished the account of our labours in establishing the southern boundary of the United States, I shall make a few observations on the prevailing diseases, that made their appearance while I was in the country which has been the subject of this journal.

The prevailing diseases on the lower part of the Ohio, on the Mississippi, and through the Floridas, are bilious fevers. They vary in their forms according to the state, or the force of their remote and exciting causes: some seasons they are little more than the common intermittents, and remittents, which prevail in the middle states; but in others they are highly malignant, and approach nearly to, if not become the genuine yellow fever of the West Indies.

Although fevers are the prevailing diseases in the tract of country above mentioned, others are not uncommon, which though very different in their appearance, probably owe their origin to the same causes. In the summer of 1797, while our people were afflicted with bilious fevers, the Spanish soldiers in the same town were suffering with dysenteries, and which generally proved mortal. During the prevalence of those two diseases, our physician reported to me two cases of genuine pleurisy, both of which yielded to bleeding. At that time, (and I yet see no reason to alter my opinion,) I supposed those different complaints had their origin from the same cause, and their different appearance arose from accidental causes, such as difference of constitution or the manner of living. Our physician, who for nearly two months attended the Spanish soldiers, informed me that medicine had but little or no effect upon those laboring under the dysentery, and which I was afterwards informed by some intelligent Spanish gentlemen, was thought to be owing to the profuse use of stimulants, such as Cayenne pepper, by their soldiers in that country, which by its inordinate use destroyed the tone of the intestines, and rendered them insensible to the most active medicines.[56]

56. Ellicott provides some very useful information at this point on the causes, symptoms, and treatments for diseases common to the area. He goes on to speculate on a number of

It is probable that the first attack in each of those complaints was inflammatory, and had bleeding been resorted to in that stage, it would have been attended with salutary effects. The late Col. Clarke, of the Mississippi Territory was in that season attacked with a most violent fever, which in a few hours almost deprived him of eye sight, and finding that no time was to be lost, and not medical aid to be had in his neighbourhood, he bled himself freely several times and recovered in a few days. The circumstance he has frequently mentioned to me, and attributed his recovery to a paper of Dr. [Benjamin] Rush's on the yellow fever, which he had read a short time before.

It did not appear to me during my residence in the country, that temperance by any man prevented the attacks of the fever, on the contrary the free livers frequently escape it, while the temperate suffer from it; but there is this difference to be observed, the temperate with good management generally recover, and on the contrary, the others when attacked commonly sink under the complaint in a few days. Several gentlemen of that class who were called free livers went into that country the same year I did; they escaped the fever two seasons, but sunk under it the third.

The natives, though not wholly exempted from those fevers, are much less subject to them than strangers. This no doubt arises from a very natural cause: the constitutions of the natives are accommodated to the climate from their infancy, while the constitutions of strangers being moulded to a different one, yield more readily to those diseases. For, although the human species can exist in all climates, the constitution appears to be naturally adapted to that in which the person is born, and raised, and therefore upon changing the climate, the constitution is generally found to change also, and this change, (which is called the seasoning,) is commonly effected by the prevailing endemic of the country. This change is very severe on the firm constitutions of our northern citizens, which like strong oaks in a tempest, are broken off or torn up by the roots, while weak constitutions like flexible reeds, yield to the tempest, and rise when the storm is over.

At Natchez, in the month of June 1797 we had a few cases of the fever

---

medical issues in not only the Southwest but also the Middle Atlantic and northern states. The age-old medical practice of "bleeding" was still a common remedy, but this treatment was supplemented by the use of local remedies and the widely used "thunderbolts" of Dr. Benjamin Rush. Ellicott's observations are consistent with those of Garrett Elliott Pendergrast, whose heretofore unpublished manuscript of 1803, prepared as dissertation for the degree of doctor of medicine, reinforces and complements the writing of Ellicott.

among our people, but the complaint was not general until about the middle of July. The attacks were then severe, and one of my assistants and several of our people were then carried off. Some of those who survived, were for several months extremely debilitated by frequent relapses, which appear to be almost unavoidable in that country: because it is to be observed, that the causes which produce a bilious fever in that climate, are of much longer continuance than in the middle and northern states, and the system consequently as longer time predisposed to yield to those causes.

As soon as our people recovered from the first attack, I had them removed about seven miles from the river in the country, and to prevent intemperance, and enforce regularity, I went and resided among them.

As those fevers frequently produce a languor, or partial lethargy which is generally removed by action, athletic exercises were encouraged, particularly playing ball, for which purpose a convenient ball-alley was prepared. The good effects of removing from the river, and the plan of exercise were soon visible in the recovery of the greater part of our people; some however had relapses, which though generally not as severe as the first attack, they appeared to increase the difficulty of removing the complaint, and several of the cases terminated in visceral obstructions which with one exception yielded to mercurial. The unfortunate person was a Mr. Hamilton, whose case ended in a confirmed dropsy under which he sunk after we returned to Philadelphia in 1800.

During the prevalence of those fevers my own health was remarkably good, which in part I attributed to the use of some pills given to me by Dr. Rush when I left Philadelphia. I began to take them in the month of May, though very sparingly, until in July when I used them twice a week, from which I found no inconvenience, and never lost a single meal by them; on the contrary my appetite was rather increased, and my flesh became more firm. Each of these pills [a purgative commonly known as "Rush's Thunderbolts"] was composed of two grains of calomel, with half a grain of gamboge, combined by means of a little soap.

About the 20th of September the pills were exhausted, and it was the general opinion that the danger of an attack for that season was past: in conformity to this opinion, I returned with our people on the 27th of the same month, to the town of Natchez. I should have delayed our return for some weeks, but from a desire to be near my apparatus, and complete a course of astronomical observations which I had begun soon after my arrival.

On the 7th day of October I became very much indisposed with shooting pains in my head, an oppression at my breast, but without any nausea, a pain [in] my bones, neck, and back, flushed countenance, with the appearance of a strong plethorick [excessive amount of blood] habit. These symptoms were quickly followed by a violent fever, which frequently rendered me delirious. From the strength and fullness of my pulse, affection of my head, and flushed countenance, I was of opinion that the commencement of the attack was highly inflammatory, and that bleeding would have been attended with salutary effects, but the strong prejudices against it in that part of the country, prevented its being done. The fever began to abate in seven or eight days, but left me so much debilitated, that with several relapses, it was some months before my recovery was complete.

It is probable that if the pills had not been exhausted, and I had continued the use of them a few weeks longer, until the cool weather commenced, I should have escaped the attack that season.

Various theories have been devised by physicians to account for those malignant fevers which are the scourge of the human race, but the most natural appears to be that of supposing a portion of the system by some cause or other, to be rendered unfit for animal life, and therefore obnoxious to the healthy part, which from a natural impulse is constantly endeavouring to expel the morbid matter, and which is probably thrown off by both the external and internal surfaces. If this theory be just, the best means of prevention appears to be gentle cathartics, and frequent cleansing the outward surface by bathing, and changing of the clothes, particularly such as come in contact with the skin. To this practice I attributed my escaping the fever during the time of its general prevalence. If the offensive matter is thrown out of the system as fast as it is generated, and removed from the surfaces, I am of the opinion that but little danger is to be apprehended; but if on the contrary, it is suffered to remain in the intestines, and on the external surface, and become entangled in foul clothes, the danger will be much increased.

The variety of diseases in our southern country is not so great as in the middle or northern states. The degree of cold in the southern states, and the Floridas, is not sufficient to make a complete change in the appearance of the prevailing diseases, or to generate those which may be considered endemic in cold countries. But it is different in the middle, and some of the northern states, which have not only a sufficient degree of heat, but such a continuance of it in some seasons, as to produce the malignant fevers of the southern states, and West Indies. The degree of cold with its continuance, is also sufficient in some winters, to produce those diseases which prevail most in cold countries, and hence a greater variety of diseases may naturally be expected.

It has been doubted by some, whether the climate of the middle, and some of the northern states, is capable of producing the malignant fevers of our southern states, and the West India islands; but those doubts would in my opinion be removed from the mind of any persons, who should reside a few years in the latter, unless he was previously wedded to a preconceived hypothesis, or supposed facts, and his mind entrammelled by prejudice.

A reluctance to admit truth, is little less injurious than the propagation of falsehoods, and the longer we contend that the climate of the middle and some of the northern states, is incapable of generating the malignant fevers of the southern states, and West Indies, the longer we shall be in danger of suffering by those scourges; for while our measures are only taken to oppose a foreign enemy, a domestic one may begin its ravages. Experience teaches us that there are generally three things necessary to the production of the malignant southern fevers, first heat, secondly water, swamps or marshes, and thirdly a collection of persons. And whenever we have a long continuation of heat, aided by the miasmata [waste] from impure water, and marshes partially dry, loaded with putrid vegetables, added to a large collection of persons, each of whom by respiration is constantly rendering one gallon of air per minute unfit for the functions of animal life, we are in danger of being attacked by a malignant fever.

From the locality of those fevers in the United States, may not a conclusion be fairly drawn, that the cause, or causes, is, or are, in some degree local also? For if this were not the case, those fevers would not be confined to our large towns on the water, but extend generally over the face of the country; which is contradicted by experience. And again, if the fever had its origin from importation, why is it confined to particular places? The answer it is presumed would be, that in those places there is a greater predisposition from some exciting cause, whatever it may be, to receive the infection. Now let us see to what point this answer would conduct us. If in those places there is a greater predisposition to receive the infection, it follows, that this predisposition in some degree depends upon local causes: and if the causes which produce this predisposition to receive the infection, can from concurring circumstances be increased, may not a just and logical conclusion be drawn, that they may be so heightened, as to produce that species of fever which in a milder form they prepare the system to receive?

As those fevers appear evidently to depend in part upon local causes, the means of prevention will in an equal degree depend upon removing, or correcting those causes.

It is the opinion of many persons, that our large commercial cities would be materially injured, if they were thought capable of producing the malignant fevers of our southern country, and the West Indies. This opinion however plausible, certainly rests upon a slender foundation, because this opinion alone cannot prevent the recurrence of those fevers, and it must be the recurrence, whatever may be the origin that will eventually be found injurious.

If those fevers can possibly be generated in our large commercial cities, in the middle and southern states, we may, as has already been observed, be attacked by a domestic enemy, while our measures are only taken to avoid a foreign one: And if it should be discovered that those fevers are not of domestic origin, it must be granted that from some cause or other, there is a greater predisposition to receive the infection in our large towns and villages, situated on our rivers, than in other places; an investigation of this cause would therefore be a subject of the highest importance, for in all probability the removing the cause would secure us against this scourge, so injurious to the interests, population, and happiness of our country.

On this part of the subject I should have been silent, had it not been to correct an opinion which I entertained some years ago, that the climate of the middle states could not generate that species of the bilious fever commonly called the yellow fever; but from a residence of several years in the southward, where those malignant fevers frequently appear, I feel a strong conviction, that in the middle, and some of the northern states, there is both a sufficient degree of heat, and continuation of it some seasons, to generate those fevers in places situated in the neighbourhood of swamps, ponds of stagnate water, and in the midst of such filth, as will too frequently be found about large cities and towns.

The reports with the maps of the boundary being completed on the 10th of April, we packed up our instruments and baggage, and on the 11th left Cumberland island, and sailed to the town of St. Mary's where we remained until our Commissary had finished his business at that place, which was on the 25th, when I again took the direction of the vessel, having but one sailor exclusive of two or three of our labourers who had come round Cape Florida with me. In the afternoon the wind serving, we left St. Mary's and proceeded along the sound to the Plumb Orchard, where we were becalmed and come to an anchor.

The 26th we were becalmed a great part of the forenoon, in the meantime I paid a visit to the family of the late Major General Green, who now resides on Cumberland island. Got under way about 11 o'clock A.M. and proceeded into St. Andrew's Sound and attempted to cross the bar, and put to sea; but having no pilot, and the passage appearing somewhat difficult, we returned and sailed into St. Simon's Sound, and came to an anchor.

27th. The wind serving about 8 o'clock in the morning we crossed the bar, put to sea, and laid our course for Tybee light-house. By a good observation found the latitude of the north end of St. Simon's island to be about 31°16′14″ N. A fresh breeze from the S. all the afternoon: about 5 o'clock P.M. spotted a schooner from Rhode-Island at anchor, she was at least 2½ leagues from land. About 9 o'clock in the evening we saw the light in the lantern of Tybee light-house. The wind almost died away, and we made but little way. A short time after midnight, the light in the lantern of the light-house went out.

28th. Very hazy in the morning, and almost calm. At 1 o'clock P.M. the mist was carried off by a smart breeze, when we found ourselves about one league from the light-house, and immediately made a signal for a pilot; but after waiting until half past 5 o'clock in the afternoon, and no pilot appearing, and the wind having become violent from the N.E. with a heavy sea, we took in the top sail, reefed the fore sail, and put to sea for fear of being driven on the coast and wrecked.

29th. Wind from the same quarter and very strong.

30th. Calm until 9 o'clock A.M. when we were favoured with a fine breeze from the S. By a good meridional observation found our latitude 31°38′47″ N. Stood north until 3 o'clock P.M. when we saw the light-house: sailed to the bar, crossed it about 6 o'clock in the evening, in company with a sloop the master of which was acquainted with the channel, and come to an anchor.

May 1st. The wind serving about 10 o'clock A.M. we sailed up to the city of Savannah, where our Commissary was employed until the 8th in settling some of his public accounts. In the mean time, the necessary arrangements were made for sending our schooner, which had been fitted up at New Orleans, round to our newly established military post on the Mobile. Three reasons offered for this measure. First, the vessel was flat built, and particularly calculated for the coasting trade on the Gulf of Mexico. Secondly the United States had no vessel for that important post, in consequence of which it was dependent on the subjects of his Catholic Majesty for the necessary supplies by water, and thirdly, the free navigation of the Mississippi alone had been secured by our late treaty with his Catholic Majesty, and no provision having been made for the uninterrupted use of the navigable rivers which rise in the United States, and fall into the Gulf of Mexico, between the Mississippi and East Florida. It was therefore expected, that by this measure, the free navigation of those waters would either be secured by usage to the citizens of the United States, or produce such an explanation as would effect that object.

Our Commissary having finished his business on the 8th, chartered a sloop belonging to New York, for our passage to Philadelphia, and at 9 o'clock in the morning the pilot came on board, and we proceeded down to the light-house.

9th. Got under way at 7 o'clock in the morning, and were accompanied over the bar by the Revenue Cutter, after which she fired a salute and left us. The master of the sloop who was also the owner, had no other instruments on board than his compass, and two old Davis's quadrants, neither of which could be depended on for the latitude nearer than 15 or 20 minutes; he expressed a wish that I would make use of my instruments, and take part in the navigation of the vessel, so far as related to the courses, distances, latitude and longitude, to which I readily agreed. Our passage was a disagreeable one, owing to contrary winds, and sudden squalls; in one of which were near losing our mast.

On the 10th we entered the Gulf Stream, and continued in it until by observation we found ourselves north of Cape Hatteras, and then stood for Cape Henlopen; but the winds were so unfavourable that we did not make land until the evening of the 16th, when a pilot came on board. Early in the morning of the 17th entered Delaware Bay, and having a good southerly breeze great part of the day, we reached Chester in the evening and come to an anchor.

18th. Got under way at 6 o'clock in the morning, the wind very light, and right a-head; come to an anchor at Fort Mifflin, and after having been examined by the Physician and Health-officer, were permitted to proceed; but the tide failing, we did not arrive at Philadelphia until 8 o'clock in the evening, when all the fatigues, hardships, and difficulties I had been exposed to during a long absence, were more than compensated by the pleasure I experience in meeting my family in good health.

I cannot take leave of the different subjects of this journal, without acknowledging my obligations to that kind Providence, which preserved me from the dangers that have been described and probably from many others unseen, and unknown, during a tedious absence from my family, friends, and native country; nor can I take a final view of a part of that extensive country I have traversed, without expressing the pleasure I enjoy in contemplating it, (in consequence of some late events) as the future theatre of free governments, and of the rapid population, and national, and individual happiness that are connected with them. May the now peaceful shores of the Mississippi, never be made vocal with the noise of the implements of war, and may its waters never be dyed with human blood!—With this wish, thanking my reader for his patience in looking over these pages, I bid him adieu.

APPENDIX

1. Excerpts from Journal of a Tour in Unsettled Parts of North America in 1796 & 1797, *by Francis Baily (1774–1844). edited by Jack D. L. Holmes, 145–57 (Carbondale: Southern Illinois University Press, 1969).*

Francis Baily was an English merchant in his early twenties when he made his tour of the United States. Traveling down the Ohio and then the Mississippi Rivers, he recorded everything he saw of interest, which included observations of events in Natchez at the same time that Ellicott was there. He was in frequent contact with Ellicott in that his business dealing with the Spanish included the attempt by Joseph Vidal, secretary of the Spanish district at Natchez, to pay his debts with depreciated paper currency. Baily therefore appealed to Ellicott for redress of his complaints per Article XX, regarding commercial relations, in the Treaty of San Lorenzo. Baily's notes provide additional details about the Old Southwest under Spanish rule, and his business dealings in Natchez and New Orleans under Spanish administration reinforce similar views expressed by Ellicott. A portion of his Journal of a Tour was published in 1844; only that portion pertaining to his Natchez observations, from May 3, 1797 to June 1, 1797, is excerpted here.

Wednesday, May 3rd—we started from this place [Chickasaw Bluffs] about nine o'clock. A great many Indians were assembled on the shore to see us depart; others had taken their guns, and were gone a hunting in the woods. The wind was rather high, and we had not floated above an hour, ere we were obliged to put ashore again. We stayed near an hour, and the wind abating a little, we floated again. At night we came to again on the signals being made, and the next day—

Thursday, May 4th—we passed the river St. Francis, about seventy miles from C. Bluffs. Our sight of its mouth was cut off by some islands, among which we were floating at the time we passed it, so that we don't know exactly the time when we came to it. This is but a small river, and rises a little way in the interior of Louisiana; and the banks just above its mouth are made a resort for hunters, who often meet here both in going out to hunt and returning with their prey. It is an old encamping ground, and on that account is made use of by the hunters to get their things ready for their journey into the country.

Saturday, May 6th—we put ashore this evening, not far from the river Arkansas. This is a considerable stream, and has its source not far from Santa Fe, in the province of New Mexico. It runs through an immensely rich and fertile country, and is said to navigable for bateaux [flat boats] for 700–800 miles. The Spaniards had a fort [Fort Carlos III] about ten or twelve miles up this river, for the purpose of defending trade carried on with the Arkansas Indians. An inundation of the Mississippi some years back caused the evacuation of this fort, and the establishment of another on the north bank, about thirty-six miles higher up. This fort is still kept up, and the Spaniards are giving great encouragement to emigrants to settle there. The Arkansas discharges itself into the Mississippi by two mouths, the upper of which is called Riviere Blanche, from its receiving a river of that name, which is said to be navigable 600 miles, and the soil through which it runs, equal in quality to any in the Mississippi. The Arkansas is about 110 miles from the river S. Francis.

Tuesday, May 9th—About half-past two o'clock in the afternoon we passed the Yazou river, about 160 miles from the Arkansas river. This stream rises high up in the Cherokee country. It runs through a very fertile soil, and empties itself into the Mississippi by a mouth about 100 yards wide. About sixteen miles up this river the French had formerly a settlement, but it was destroyed by the Yazou [Natchez] Indians in 1726 [1729]. This tribe of Indians is now entirely extinct. This is the river which has been the bone of contention between the United States and the Spaniards: the latter claiming the country to southward of it as being included in the province of West Florida, of which the Yazou was the northern boundary; and the Americans maintaining, on the other hand, that the northern boundary of West Florida was the 31st degree of north latitude. A great deal might be said in support of both claims, though I think most in favor of Spain; however, as by the late treaty [1795] the Spaniards have agreed to give it up, it will be needless to enter into an unprofitable discussion of its merits.

About five o'clock we came to the Walnut Hills, called the Yazou Cliffs, in Hutchins's Map, twelve miles from the Yazou river. Here there is a strong fort [Los Nogales] kept up by the Spaniards. It is an irregular fortification, occupying a great part of the hill on which it stands, which is very high and steep. Here we ought to have put ashore to show our passport. But though we submitted to this degradation in going ashore at the forts established on the Spanish territory, yet at this place (which was a fort established within the American lines), and unlawfully kept possession of by the Spaniards, in contradiction to the treaty lately [1795] concluded between the two countries we were determined to assert our own rights, and not comply with so unjust and humiliating a demand: accordingly, we floated by without taking any notice of them; and we had scarcely got opposite to the fort, ere we had a gun fired at us, which was a signal for us to heave to; but we, regardless of their threats, continued on, and by the rapidity of the stream were wafted out of their sight ere they could load another piece to bear upon us. The other boats in our company who were behind, fearful that they should pay for our contempt of the summons, obeyed the signal, and rather than run the gauntlet of their pieces, put ashore. We put ashore a little below this place, and were soon joined by another of the boats which had made its escape from under the fort.

Walnut Hills is a beautiful situation for a town, and an advantageous one for a fort. There are two forts at this place, one of them commands the other, being situated upon an eminence behind it. The few houses which are scattered around it, and the green bank on which they stand, surround with flowering, verdant, and lofty trees, presented at once a picturesque and romantic appearance to our eyes, fatigued with the uniformity of the prospect to which we had for so many miles been witness.

The weather being very fine, and the moon shining very bright, and there being very few sawyers [vegetation embedded in the river bottom] in the river just here, we determined upon proceeding on our journey, in order that we might get to the Grand Gulf [confluence of the Big Black and Mississippi Rivers] by the middle of the day, as being the most proper time for passing that dangerous spot. Accordingly, having taken each of us a nap, we got up about half-past twelve, and having got our boats into the middle of the stream, one part of our company kept their appointed watch, and the other retired again to rest, and the same morning, about eleven o'clock—

Wednesday, May 10th—we came to this perilous vortex, which is the most dangerous place in the whole navigation of the Mississippi. The river here is thrown up with great impetuosity against the bluff point of a rock, which opposes its broad course in numberless whirlpools, into one of which if a boat gets, she is carried round with an astonishing rapidity, like a whirligig, and becomes unmanageable; so that if the direction of the vortex happens to be towards the rock, she must inevitably be dashed to pieces. The river then makes a very sharp turning round a point of land directly opposite the rock, and runs a course immediately contrary to the one it before pursued. The way to escape this place, and pass in safety through its terrors, is to keep the boat exactly in the middle between the current which runs towards the rock, and the eddy or countercurrent which runs near the point; for, in all these places there is a counter-current [that] runs along the opposite shore; into which if you happen to get, you are carried back, and have to go through the same trials and difficulties a second time. This difficulty, then, we endeavoured to surmount; and to do it required our constant presence at the oars, and a steady attention to the commands of the person who undertakes to conduct her, and who is upon the top of the boat, observing the course of the numerous currents. Happily, two of our company had passed it before, and were, therefore, somewhat acquainted with its navigation. To them we consequently entrusted the management; and by their steady attention, we had no sooner arrived at the critical point, than we shot through it all in safety, like an arrow from a bow, or like a body precipitated from a mill-tail. Grand Gulf is about fifty miles from Walnut Hills.

At one o'clock, P.M. we came to Bayou Pierre [Stoney River]. This is a little stream which rises up in the district of Natchez; and upon the head waters of which, there are some settlements, which form part of that district; there were also two or three plantations at its mouth. Here we went ashore in our canoe, and got some eggs and milk, which were acceptable to us who had been so long deprived of every luxury of this kind. The land here was very nearly overflowed, being very few inches above the level of the river. The inhabitants told me that they never remembered the river so high. We did not stop here many minutes, as our boat passed by as swift as lightning, and we were obliged to make the best of our way to catch her before she arrived at the Little Gulf, which is a place, in its situation and effect, somewhat like the Grand Gulf, only on a smaller scale. Its danger also is scarcely any in high water; but when the river is low there are some strong eddies which ought to be avoided. It is about ten or twelve miles from Bayou Pierre. The next morning—

Thursday, May 11th—about twelve o'clock we arrived at Natchez, sixty miles from Bayou Pierre. This is the capital of the district which goes under this name. It is situated upon a high hill, which terminates in a bluff at the river, consists of about eighty or ninety houses scattered over a great space of land. The streets are laid out upon a regular plan; but there is so much ground between most of the houses, that it appears as if each dwelling was furnished with a plantation. There is a fort upon an eminence near the river, which commands both the town and the Mississippi; but it is in a ruinous condition, and could not be defended against a regular attack. The governor of this place is Don Manuel Gayoso de Lemos, a man who (for a Spaniard) is said to have behaved tolerably well in his office; but let everyone speak well of the bridge which carries him safe over. On our approach to the shore, we had the pleasure of beholding the American colors flying on the banks. Agreeably to the treaty entered into between Spain and America, the former agreed to evacuate and give up all the country to the eastward of the Mississippi, which was to the north of the 31st degree of north latitude, in which the district of Natchez is included; and the commissioners for determining the precise point where the 31st degree of latitude commenced, and for running the line which was to be the boundary between the two countries, were (agreeably to the treaty) to meet at Natchez; and as soon as it was ascertained what forts were to the northward of that line, they were to be evacuated. Mr. Ellicott, the commissioner on the part of America, had been here ever since; and as it was pretty well known that Natchez would fall to the side of the Americans, Lieutenant [Piercy Smith] Pope was sent down with a party of men to take possession of the fort as soon as the Spaniards should evacuate it. The latter had, some little while back, shown a disposition to give them up, and had actually removed several of the cannon down to the boats appointed for their reception, both at this fort and at the Walnut Hills; (which was still more to the northward;) and they had even destroyed the fort at Chickasaw Bluffs, as I have before mentioned. But on the appearance of a rupture [1797] between the United States and France, (with the latter of whom Spain was in alliance,) they suddenly countermanded these orders, and the governors were instructed to replace the cannon, and put the garrison in a posture of defense, and not deliver up the forts. [Ellicott reported this development to Secretary of State Pickering on May 27, 1797.] It was under this aspect of things that Mr. Ellicott and Lieutenant Pope arrived; and instead of finding an amicable disposition in the governor [Gayoso] to give up the fort, and proceed to the determining of the line, he absolutely refused the former, and unnecessary delays protracted the latter. However, as Lieutenant Pope could not go counter to his orders, he landed his men, and marching them up the hill, took possession of an eminence immediately opposite the fort, and there, hoisting the American flag, he encamped his men; and it was in this situation that I found him when I was introduced to him by Mr. Ellicott, whom I had known at Pittsburgh. He informed me that Gayoso sent to him, soon after he had been there, to strike his colors, saying, he had no right to hoist them on the Spanish territory. On his refusing so to do, he threatened to send and cut down the flagstaff; but on Lieutenant Pope's assuring him that he would defend his colors to the utmost, Gayoso gave up the contest.

This district has been settled principally by English and Americans; and though the country was given up to the Spaniards in 1783 [1784], the proportion of Spanish inhabitants is very small. To persons brought up under a form of government to which the English and Americans have been accustomed, the Spanish government must be an intolerable yoke. They depend in all their civil and criminal affairs upon the whim or caprice favour or folly of an upstart Spaniard who is set over them as their governor, and who, through pique or malice, or in a fit of drunkenness or insanity, has it in his power to sport with the lives and property of those persons over whom he is placed for the ostensible purpose of protection. Abusing this trust in the most shameful and despotic manner, as they often do, even in hazard of the safety of the inhabitants, it is no wonder that the people composing the district received with pleasure the news of the territory being delivered up to the Americans, and that they should soon get rid of their haughty master, under whom they had suffered so many hardships and inconveniences; and that they saw, with regret, mixed with the greatest resentment, a disposition on the part of Spain to violate the treaty, and not deliver up the fort, together with the country. This just resentment was carried to a great pitch whilst I was here, and broke out in open acts of violence several times; and at last proceeded so far as to induce the governor to retire into the fort [June 1797], and to call upon all the people attached to his person to come to him, and defend themselves against the designs of those evil-disposed men (as he denominated them). There were about a dozen flocked to his standard; as to the rest of the district, they surrounded the fort, and kept his Excellency prisoner there near a fortnight, and would not let him come out at last, till he had signed articles of capitulation; which articles are as in the annexed note, and clearly indicate how determined the inhabitants were to maintain their just rights. To particularize all the incidents which took place at this period relative to this subject would fill a volume; they are all sufficiently related in the reports of the Secretary of State to the Congress.

The town of Natchez is situated in north latitude 31°33′46″, and west longitude 6°5′57″ from Greenwich: the whole district may contain about 5,000 inhabitants. The houses are chiefly framed buildings; but, though this country has been settled so long, there is all that inattention to neatness, cleanliness, and the comforts attending thereon, that there is in a country just cleared. I have seen houses in this place (and those possessed by persons assuming a degree of consequence in the country) scarcely furnished beyond the first stage of civilization, when a few boards nailed together have served for a bedstead, and a mattress covered with a few blankets for a bed, when there has been scarcely a chair to sit down upon, or a table to place anything on, but everything in the greatest confusion and disorder about the room. This, to be sure, is not universally so: on the contrary, I have seen others fitted up in the neatest manner possible; but then the greatest plainness, without any of those luxuries which decorate even the cottages of our English farmers. The climate is delightful, though in the summer I think somewhat too warm, the thermometer being here in June as high as 107°. Ice is not known here, and snow but seldom, and then very thin, and soon goes off. Its situation is pleasant, being the uneven surface of the top of a high hill, which commands a fine view of the Mississippi for a considerable way, as well as of the country. From the point next the river you may look upon the borders of the water, and see the alligator prowling along amongst the bushes and brambles which are in the bottom, and at times uttering the most dismal howlings. These animals, whose hides are impervious by a musket ball, are sometimes caught by the Indians by a manoeuvre truly his own:—He goes armed with a strong hickory stick, about two feet long, barbed at each end, and which he holds in the middle as tight as possible. In the other hand he takes some article of food to attract them, and to induce them to open their enormous mouths to obtain it, and which the Indian holds out to them; but no sooner does the alligator make the attempt to seize it, than the Indian snatches that arm away, and presents the other furnished with the double dart. The alligator, unconscious of this, closes his mouth upon his supposed prey; and unable to extricate himself or open his jaws, the Indian drags him to shore, amidst the applause and acclamations of the spectators who stand by admiring the daring act. The roads about here are very good, considering there is no attention paid to them; the usual mode of travelling is on horseback; and as there are no public-houses, a spirit of hospitality is kept up between all neighbors. This hospitality, which is only shown amongst neighbours, or the friends of neighbours, I shall more fully treat upon, when I come to touch upon the manners and customs of the Americans in general; (for Americans I still consider these people;) though I shall touch upon the state of society and mode of living when I come to take leave of the district. There is a great deal of cotton raised in this district, which is sent down the river to New Orleans: it is of the nature of Georgian cotton. There are several jennies [common term for the cotton gin, which was introduced here about 1795–96] erected in the neighbourhood, in order to extricate the seed from the cotton. There is one immediately in the town on the banks of the river, belonging to [Stephen] Minor and [Robert] Scott, worked by two horses, which will give 500 lbs. of clear cotton in a day. They have one-eighth part for their trouble. The seed-cotton loses three-fourths of its weight by jenning. Very good tobacco and rice is raised here, but in no considerable quantity.

There are two or three places here which go under the denomination of Taverns, and where you may get accommodated with board and lodging. I put up at one of them, (at which there was a billiard table kept) and paid my landlord a dollar per day, which was enormous, considering the fare; for provisions are not very plenty in this province, at least, if we judge from the prices. Imported articles must come high; but I think it possible that their markets might be better supplied than they are; in fact, I have no doubt but they will when the American mode of government comes to be administered, and the persons and property of the inhabitants to be protected, and full encouragement given to industry in all its forms. Looking forward to this time, we may pronounce this district to be the most flourishing in the southwest territory; and the town of Natchez for to excel every other on the banks of the Mississippi. The land around is of an excellent quality; and by the near port of New Orleans, it has an easy means of exporting its produce, and receiving in return such articles of foreign manufacture as may be most in demand. Land in the country is sold for about a dollar an acre: a five-acre lot close to the town sold for 150 dollars. Dr. Watrous (our fellow passenger) bought a lot of 150 acres of un-cleared land near the town for four dollars per acre, and it was thought cheap. The article of land must nevertheless depend in its value upon its relative situation and advantages, as well as upon its quality. Upon the whole, I think this an excellent place for a person to settle in, (when it comes under the government of the United States,) if he can bring himself to give up the advantages of refined society; though I don't know that this remark is more particularly needful here than in all young countries: on the contrary, I know several persons here, both Spanish and English, whose conversation and company have been interesting and amusing. Slavery is permitted by the Spanish government, and no doubt will be continued by the Americans, till they have adopted some measure for the utter annihilation of it from the country.

Of the Spanish government in itself I shall make no remark till I relate all together my opinion on it from New Orleans. I shall mention a circumstance relating to myself, which set in a strong point of view the oppressive and domineering conduct of the Spanish governors, as well as inform you of the hazardous situation I have been in.

The secretary of the government (one Joseph Vidal, a Spaniard) had purchased of me the remainder of those goods which I had, after trading with the Indians, amount to about 680 pesos; these he was to pay me for immediately, but when I called on him for the money, he offered me in payment a certificate to nearly the amount. These certificates are a species of paper-money drawn by the commanders of the different forts on this river on the treasury at New Orleans, and are paid away to workmen, soldiers, etc., instead of money, and are received by the merchants as such; so that they are a kind of bank notes received upon the faith of government, though it must be observed that oftentimes there is no great sum of money in the treasury, and these certificates are returned unpaid, so that they generally bear a discount: and this discount is proportioned to the degree of confidence put in the prospect of getting the money. It happened that they bore at this time a discount of 12 per cent., and yet this unreasonable rascal wanted me to take them at their full value, which I of course, refused, and wished him to pay me in cash. He at first seemed to hesitate, and said that he would try if he could get it discounted; but on my calling again, said he was under the necessity of telling me I must take it as cash; and as I found all means to induce him to the contrary were of no avail, I appealed to the governor

[Gayoso]. I have before hinted that the governors are all, directly or indirectly, concerned in a contraband trade, and I had every reason to believe that he was connected with Vidal, in this instance; I therefore did not much flatter myself with the prospect of success. However, I went to him, and told him my tale: as a boy at school would go the master, and complain of the improper conduct of any of his companions, and if the master thought proper he would punish him; otherwise, he would dismiss him with impunity. Gayoso listened to me very patiently, and as he had heard it all from Vidal prior to my relation, he had made up his mind what decision to give; he said that the certificates were a legal tender, and that I could not refuse them. I was surprised to hear this notorious falsehood from a person so high in office; and finding that there was no prospect of obtaining justice here, I told him my determination of carrying the case before [Francisco Luis Héctor] Baron de Carondelet at New Orleans, who is the Commander-in-Chief of the province of Louisiana [1792–97], and for this purpose wished him to give me his decision in writing, with his hand and seal annexed. At this he seemed very angry, and threatened me with what he would do if I made an improper use of the papers, or went to misrepresent them. I soon appeased him, or at least apparently so, when I set forth the justice of my claim, and my indifference about his anger; and I afterwards asked him how I might obtain his decision. He told me that the formal way was to draw up a state of the case in the manner of a petition praying for redress, and that he would write underneath his decree. Accordingly I drew up a paper, and delivered it to him. He did not seem to like the contents of it; for though I had taken care to flatter his vanity by some expressions in it, yet he evidently saw that the case was stated too clearly to admit of a doubt of the justice of it; and, I believe, he was almost ashamed to annex so shameful and illegal a decision to it as he has done; particularly when he understood that I had consulted Mr. Ellicott and Lieutenant Pope on my plan, and that they had determined to support me in my claim, at least, so far as regarded the carrying that part of the treaty into effect which I had there claimed. Finding no notice taken, in his decision, of the law which made these certificates a legal tender, I went to him and asked him to point out, and show me the law by which he was guided; and I never shall forget the looks of the man at this (what he called impertinent) question; for, wondering at my assurance, and threatening me with the horrors of the Callibouse [jail] if I any longer disputed his authority, he laid his hand upon my breast and told me that he was the law; and that as he said the case was to be determined.[1]

When he made this speech, at which he seemed more than every enraged; and, I believe, had it not been for the neighboring situation of the American commissioner and commander, together with the general revolting spirit of the district, that I should have been hurried off to immediate imprisonment, if not to the mines. The anger of a Spaniard is so implacable and malicious, that he will leave no stone unturned to accomplish his revenge, even to the act of assassination. Under this idea, and by the advice of my friends here, who had known instances of their hateful temper, I always went guarded, and at night never slept but with a pair of pistols under my pillow; for as my door faced the road, it was an easy thing to break it open, and (hurrying me down to the river) to elude all search which have been made for me. However, this did not deter me from pursuing my cause, for I was determined on having justice if it was to be found in the country; accordingly, I went to the governor once more, and told him that I wished to protest against receiving the certificates. He said that he could not enter any protest against the king's money, as he called it. I then told him that I wished Vidal should pay the money in his presence, and that I had appointed Mr. Elliott and two other witnesses to be present at the transaction. At this he began to grow angry, and told me I must not pretend to dictate to him what to do; and that he should suffer nothing of the kind. However, I pursued my claim in due form and order, and sent in a petition to see whether he would have the assurance to deny me this just and reasonable request. I was present when it was de-

1. Spanish officials' practice of using depreciated local paper currency to pay their debts was commonplace, and decried by American merchants. See "Southern Boundary," vol. 1. On Spanish fiscal policy, James Pitot wrote: "The heads of government and their assistants have squandered the colony's funds. One of them destroyed confidence in the paper money which he gave in payment of his obligations, and which he acquired with his own silver and sometimes even that of the king. Another has made some deals in which the contracting party gave back a large part of the agreed price; and, finally, all excelled only in the thousand and one ways of emptying the public treasury in order that, the use of the local paper money making it seem legal, the possible savings passed with impunity to their own profit" (James Pitot, *Observations on the Colony of Louisiana from 1796 to 1802,* ed. and with foreword by Robert D. Bush (Baton Rouge: Louisiana State University Press, 1979), 22. See also the analysis of the depreciation of paper money in John G. Clark, *New Orleans 1718–1812: An Economic History* (Baton Rouge: Louisiana State University Press, 1970), 267–69.

livered in, and heard him mutter something about the stubbornness of the American character. In that petition I prayed that Vidal might be ordered to discharge his debt before the governor, and that three witnesses of my own appointment might be present thereto, agreeably to the provision [Article XX] made in the treaty. However, as soon as the petition was read to him, he called in three of the officers of his household together with Vidal; and after explaining the nature of the business to them, opened a kind of court for the decision of the case. I asked him if the witnesses I had nominated were not to be present; and he, answering in the negative, addressed himself to me, and asked me whether I was willing and ready to receive the money. Upon which I turned to him, and said, "Sir, you are making a mockery of justice; I shall not answer you till my witnesses are admitted:" and immediately quitted the court, leaving them to brood over their own iniquity, and to stare with stupid astonishment upon each other. The next day I called to see what they had done with my petition, and found the governor's refusal of the prayer of it written at full length at the bottom, together with a minute of my having so precipitately left the court. I immediately communicated the contents to Mr. Ellicott, and he assured me that he would send the particulars in his next dispatched to the secretary of [state] of the United States; and inform him of with how little ceremony the Spaniards treat the American citizens; and with what indifference they can break the most solemn treaties.

The boat which brought me down here having sold all its flour at this place, and proprietors intending to return to their own homes through the wilderness, I was obliged to look out for another conveyance to take me to New Orleans which is about 300 miles down the river; and from that place I had no doubt but that I should meet with a vessel to take me round to New York. This circumstance detained me here till

Thursday, June 1st—when a boat laden with cotton (among which was some belonging to myself, which I had purchased here) being ready to go down, I waited on my old friend the governor once more, in order to get my passport to proceed to New Orleans. I had been advised by several of my friends, not to trust myself any farther into the Spanish territories; they assuring me that there was no doubt but that Gayoso had represented the whole of my conduct in the strongest colours possible to the Baron de Carondelet; and that I might be taken by surprise there, when I should be deprived of the aid and support of the American commissioner and commander, to whom I have great reason to think that I owe my personal safety. I communicated these suggestions to Mr. Ellicott, and he allowed the full force of them, but at the same time assured me, that in case of any attack upon my person, he would hold Vidal as a hostage till I was safely returned.

Under this confidence I applied to the governor for my passport, which he immediately made out, glad enough (I believe) to get rid of so troublesome a visitor.[2]

2. Governor Gayoso's response and decision regarding Baily's "prayer" was that: "Certificates on the Royal Treasury at New-Orleans are considered throughout this Province as cash, and no one can refuse to receive them unless it was expressly convenanted [a written legal document] this payment should be made in silver or gold . . . ; to do otherwise would be overthrowing the practice established not only in this place but in others of greater note, such as Cadiz and other Cities in the Dominions of his Majesty" (qtd. in Jack D. L. Holmes, *Gayoso, The Life of a Spanish Governor in the Mississippi Valley 1789–1799* [Gloucester, MA: Peter Smith, 1968], 66). The situation was made more difficult for Baily because Governor-General Carondelet, to whom he wished Clark to appeal his case, was appointed to the Audiencia of Quito in Ecuador in 1796. Clark, a New Orleans merchant, had dealt with Spanish fiscal and monetary policies for years and needed to remain on the good side of the top Spanish official for Louisiana. This was Gayoso, who replaced Carondolet, after August 1797! Clark's decision was therefore based upon the realities of doing business in the Spanish provinces, not on the mistake made by an English merchant in trading with one of them (Baily, *Journal,* 184–85).

SELECT BIBLIOGRAPHY

ARTICLES

Alexander, Sally Kennedy. "A Sketch of the Life of Major Andrew Ellicott." In "Records of the Columbia Historical Society" (bound archival volume), 158–202. Washington, DC, 1899.

Arena, C. Richard. "Land Settlement Policies and Practices in Spanish Louisiana." In *The Spanish in the Mississippi Valley 1762–1804,* edited by John Francis McDermott, 51–60. Urbana: University of Illinois Press, 1974.

Bartlett, G. Hunter. "Andrew and Joseph Ellicott, the Plans of Washington City and the Village of Buffalo and Some of the Persons Concerned." *Publications of the Buffalo Historical Society* 26 (1922): 1–48.

Clarke, Agnes Mary. "Francis Bailey." In *Dictionary of National Biography,* 22 vols., edited by Sir Leslie Stephen and Sir Sidney Lee, 1:899–904. Oxford: Oxford University Press, 1882–1949.

Cox, Isaac Joslin. ed. "Documents on the Blount Conspiracy, 1795–1797." *American Historical Review* 10 (1905): 574–606.

Din, Gilbert C. "Spanish Immigration Policy in Louisiana and the American Penetration, 1792–1803." *Southwestern Historical Quarterly* 76 (1973): 255–76.

Douglass, Elisha. "The Adventurer Bowles." *William and Mary Quarterly,* 2nd. ser., 6 (1949): 3–23.

Gallalee, Jack D. "Andrew Ellicott and the Ellicott Stone." *Alabama Review* 18, no. 2 (April 1965): 92–105.

Grant, Ethan. "The Treaty of San Lorenzo and Manifest Destiny." *Gulf Coast Historical Review* 12, no. 2 (1997): 44–57.

Harkins, John E. "Legal and Judicial Aspects in Spanish Louisiana." In *Readings in Louisiana History,* edited by Glenn R. Conrad, 41–48. New Orleans: Louisiana Historical Association, 1978.

Hatfield, Joseph T. "William C. C. Claiborne, Congress, and Republicanism, 1797–1804." *Tennessee Historical Quarterly* 24, no. 2 (1965): 156–80.

Haynes, Robert V. "Territorial Mississippi." *Journal of Mississippi History* 64, no. 4 (2002): 283–305.

Holmes, Jack D. L. "The Choctaws in 1795." *Alabama Historical Quarterly* 30, no.1 (Spring 1968): 533–49.

——. "Fort Ferdinand of the Bluffs: Life on the Spanish-American Frontier, 1795–1797." *Papers of the West Tennessee Historical Society* 13 (1959): 38–54.

——. "Law and Order in Spanish Natchez, 1781–1798." *Journal of Mississippi History* 25, no. 3 (July 1963): 186–201.

——. "Maps, Plans, and Charts of Louisiana in Spanish and Cuban Archives: A Checklist." *Louisiana Studies* 2, no. 4 (Winter 1963): 183–203.

——. "Some Economic Problems of Spanish Governors in Louisiana." In *Readings in Louisiana History,* edited by Glenn R. Conrad, 55–60. New Orleans: Louisiana Historical Association, 1978.

——. "The Southern Boundary Commission, the Chattahoochee River, and the Florida Seminoles, 1799." *Florida Historical Quarterly* 44, no. 4 (1966): 312–41.

——. "Spanish-American Rivalry over the Chickasaw Bluffs, 1780–1795." *Publications of the East Tennessee Historical Society,* no. 34 (1962): 26–57.

——. "Stephen Minor: Natchez Pioneer." *Journal of Mississippi History* 42, no. 1 (1980): 17–26.

——. "William Dunbar's Correspondence on the Southern Boundary of Mississippi, 1798." *Journal of Mississippi History* 27, no. 2 (1965): 187–90.

Kinnaird, Lawrence. "The Significance of William Augustus Bowles' Seizure of Panton's Apalachee Store in 1792." *Florida Historical Quarterly* 9 (1931): 156–92.

Liljegren, Ernest R. "Jacobinism in Spanish Louisiana, 1792–1797." *Louisiana Historical Quarterly* 22 (January 1939): 47–97.

Lowry, Charles D. "The Great Migration to the Mississippi Territory, 1798–1819." *Journal of Mississippi History* 30, no. 3 (1968): 173–92.

Nasatir, A. P. "The Anglo-Spanish Frontier on the Upper Mississippi, 1786–1796." *Iowa Journal of History and Politics* 29 (1931): 155–232.

——. "Anglo-Spanish Rivalry on the Upper Missouri." *Mississippi Valley Historical Review* (December 1929–March 1930): 359–82, 507–28.

Nasatir, A. P., and Ernest P Liljegren, eds. "Materials Relating to the History of the Mississippi Valley from the Minutes of the [Spanish] Supreme Councils of State, 1787–1797." *Louisiana Historical Quarterly* 21, no. 1 (January 1938): 5–75.

Riley, Franklin L. "Spanish Policy in Mississippi after the Treaty of San Lorenzo." *Publications of the Mississippi Historical Society* 1 (1898): 50–66.

——. "Transition from Spanish to American Rule in Mississippi." *Publications of the Mississippi Historical Society* 3 (1900): 261–311.

Turner, Frederick Jackson, ed. "Documents on the Blount Conspiracy, 1795–1797." *American Historical Review* 10 (April 1903): 574–606.

Whitaker, Arthur P. "Harry Innes and Spanish Intrigues, 1794–1795." *Mississippi Valley Historical Review* 15 (September 1928): 236–48.

——. "James Wilkinson's First Descent to New Orleans in 1787." *Hispanic American Historical Review* 8, no. 1 (February 1928): 82–97.

——. "New Light on the Treaty of San Lorenzo: An Essay in Historical Criticism." *Mississippi Valley Historical Review* (March 1929): 435–54.

——. "The Retrocession of Louisiana in Spanish Policy." *American Historical Review* 39 (April 1934): 454–76.

——. "Spain and the Retrocession of Louisiana." *American Historical Review* 39 (1934): 454–76.

Young, Raymond A. "Pinckney's Treaty—A New Perspective." *Hispanic American Historical Review* 43 (November 1963): 526–35.

BOOKS AND OTHER SOURCES

Abernethy, Thomas P. *The Burr Conspiracy*. New York: Oxford University Press, 1970.

Baily, Francis. *Journal of a Tour in Unsettled Parts of North America in 1796 & 1797*. Edited by Jack D. L. Holmes. Carbondale: Southern Illinois University Press, 1969.

Bemis, Samuel Flagg. *Jay's Treaty: A Study in Commerce and Diplomacy*, New Haven, CT: Yale University Press, 1923.

——. *Pinckney's Treaty: America's Advantage from Europe's Distress, 1783-1800*. 1921. Rev. ed. New Haven, CT: Yale University Press, 1965.

Blumenthal, Henry. *France and the United States: Their Diplomatic Relations, 1789–1914*. New York: Norton, 1970.

Bolton, Herbert E. *The Spanish Borderlands: A Chronicle of Old Florida and the Southwest*. New Haven, CT: Yale University Press, 1921.

Bowles, William Augustus. *Authentic Memoirs of William Augustus Bowles, Esquire, Ambassador from the United Nations of Creeks and Cherokees to the Court of London*. London: R. Foulder, 1791.

Bowling, Kenneth R. Peter. *Charles L'Enfant: Vision, Honor, and Male Friendship in the Early American Republic*. Washington, DC: George Washington University Press, 2002.

Bush, Robert D. *The Louisiana Purchase: A Global Context*. New York: Taylor and Francis, 2014.

Carter, Clarence Edwin, comp. and ed. *The Territorial Papers of the United States*. Vols. 5 and 9, *The Territory of Mississippi 1798-1817* and *The Territory of Orleans, 1803-1812*. Washington, DC: U.S. Government Printing Office, 1949.

Caughey, John Walton. *Bernardo de Galvez in Louisiana, 1776–1783.* Berkeley: University of California Press, 1934.

Clark, John G. *New Orleans 1718–1812: An Economic History.* Baton Rouge: Louisiana State University Press, 1970.

Clark, Thomas D., ed. *Travels in the Old South: A Bibliography.* Vol. 2, *The Expanding South, 1750–1825: The Ohio Valley and the Cotton Frontier.* Norman: University of Oklahoma Press, 1956.

Clark, Thomas D., and John D. W. Guice. *Frontiers in Conflict: The Old Southwest, 1795–1830.* Albuquerque: University of New Mexico Press, 1989.

Clayton, James D. *Antebellum Natchez.* Baton Rouge: Louisiana State University Press, 1968.

Cobbett, William [Peter]. *Porcupine's Works: Containing Various Writings and Selections, Exhibiting a Faithful Picture of the United States of America; of Their Governments, Laws, Politics, and Resources; and Military Men; and of the Customs, Manners, Morals, Religion, Virtues, and Vices of the People: Comprising Also a Complete Series of Historical Documents and Remarks, from the End of the War, in 1783, to the Election of the President [Jefferson], in March, 1801.* 12 vols. London: Printed for Cobbett and Morgan, at the Crown and Miter, Pall Mall, 1801.

Coker, William S. *Indian Traders of the Southeastern Spanish Borderlands: Panton, Leslie and Company and John Forbes and Company, 1783–1847.* Gainesville: University Press of Florida, 1986.

Coker, William S., and Robert R. Rea, eds. *Anglo-Spanish Confrontation on the Gulf Coast during the American Revolution.* Pensacola, FL: Gulf Coast History and Humanities Conference, 1982.

Conrad, Glenn R., ed. *Readings in Louisiana History.* New Orleans: Louisiana Historical Association, 1978.

Cox, Isaac Joslin. *The West Florida Controversy, 1798–1813.* Baltimore: John Hopkins University Press, 1918.

Cunningham, Noble E., Jr. *The Jeffersonians in Power: Party Operations, 1801–1807.* Chapel Hill: University of North Carolina Press, 1963.

DeConde, Alexander. *Entangling Alliance—Politics and Diplomacy under George Washington.* Durham, NC: Duke University Press, 1958.

——. *The Quasi-War—Politics and Diplomacy of the Undeclared War with France, 1797–1801.* New York: Scribner's, 1966.

DeRosier, Arthur H., Jr. *William Dunbar: Scientific Pioneer of the Old Southwest.* Lexington: University Press of Kentucky, 2007.

Din, Gilbert C. *War on the Gulf Coast: The Spanish Fight against William Augustus Bowles.* Gainesville: University Press of Florida, 2012.

Ellicott, Andrew. *The Journal of Andrew Ellicott, Late Commissioner on Behalf of the United States during the Year 1796, the Years 1797,1798,1799 and Part of the*

Year 1800: For Determining the Boundary between the United States and the Possessions of His Catholic Majesty in America. Containing Six Maps. Philadelphia: William Fry, 1814.

Ethridge, Robbie, *Creek Country: The Creek Indians and Their World.* Chapel Hill: University of North Carolina Press, 2003.

Gayarre, Charles E. *History of Louisiana: The Spanish Domination.* 2nd. ed. Vol. 3. New Orleans: James A Gresham, 1879.

Godoy, Don Manuel de. *Memoirs of the Prince of the Peace.* Translated from the Spanish by J. B. D'Esmenard. 2 vols. London: Richard Bentley, 1836.

Hatfield, Joseph T. *William Claiborne: Jeffersonian Centurian in the American Southwest.* Louisiana History Series. Lafayette: University of Southwestern Louisiana Press, 1976.

Haynes, Robert Vaughn. *The Mississippi Territory and the Southwest Frontier, 1795–1817.* Lexington: University Press of Kentucky, 2010.

Henri, Florette. *The Southern Indians and Benjamin Hawkins, 1796–1816.* Norman: University of Oklahoma Press, 1986.

Holmes, Jack D. L. *Gayoso: The Life of a Spanish Governor in the Mississippi Valley 1789–1799.* Gloucester, MA: Peter Smith, 1968.

———. *A Guide to Spanish Louisiana, 1762–1804.* New Orleans: Laborde, 1970.

Hutchins, Thomas. *An Historical Narrative and Topographical Description of Louisiana and West Florida . . .* Philadelphia: Thomas Hutchins, 1784.

Kite, Elizabeth S. *L'Enfant and Washington, 1791–1792.* Baltimore: John Hopkins University Press, 1929.

Laussat, Pierre Clement. *Memoires of My Life.* Translated by Agnes Josephine Pastwa. Edited with a foreword by Robert D. Bush. Baton Rouge: Louisiana State University Press, 2003.

Linklater, Andro. *An Artist in Treason: The Extraordinary Double Life of General James Wilkinson.* New York: Walker, 2009.

Le Page du Pratz, Antoine-Simone de. *The History of Louisiana.* 1758. 3 vols. Edited by Joseph G. Tregle Jr. Published for the Louisiana American Bicentennial Commission from the 1774 English edition. Baton Rouge: Louisiana State University Press, 1975.

Matthews, Catherine Van Cortlandt. *Andrew Ellicott: His Life and Letters.* New York: Grafton Press, 1908. Facsimile of the 1908 edition. Foreword by Pat Hutchinson. Whitefish, MT: Kessinger Legacy Reprints, 2011.

McDermott, John Francis, ed. *The Spanish in the Mississippi Valley 1762–1804.* Urbana: University of Illinois Press, 1974.

McLemore, R. A., ed. *History of Mississippi.* 2 vols. Hattiesburg: University and College Press of Mississippi, 1973.

McMichael, Andrew. *Atlantic Loyalties: Americans in Spanish West Florida, 1785–1810.* Athens: University of Georgia Press, 2008.

Narrett, David. *Adventurism and Empire: The Struggle for Mastery in the Louisiana-Florida Borderlands, 1762–1803.* Chapel Hill: University of North Carolina Press, 2015.

O'Brien, Greg, *Choctaws in a Revolutionary Age, 1750–1830.* Indians of the Southeast. Lincoln: University of Nebraska Press, 2005.

Peterson, Merrill D., *Thomas Jefferson and the New Nation: A Biography.* New York: Oxford University Press, 1970.

Pitot, James. *Observations on the Colony of Louisiana from 1796 to 1802.* Translated from the French by Henry C. Pitot. Edited with a foreword by Robert D. Bush. Baton Rouge: Louisiana State University Press, 1979.

"Records Relating to the Southern Boundary of the United States." 3 vols. Compiled by Daniel T. Goggin. Record Group 76. National Archives, Washington, DC, 1968.

Robertson, James Alexander. *Louisiana under the Rule of Spain, France and the United States.* 2 vols. Cleveland: Arthur H. Clark, 1911.

Saunt, Claudio. *A New Order of Things: Property, Power, and the Transformation of the Creek Indians, 1733–1816.* Studies in North American Indian History. Cambridge: Cambridge University Press, 1999.

Smith, F. Todd. *Louisiana and the Gulf South Frontier, 1500–1821.* Baton Rouge: Louisiana State University Press, 2014.

Smith, Gene Allen, and Sylvia L. Hilton, eds. *Nexus of Empire: Negotiating Loyalty and Identity in the Revolutionary Borderlands, 1760s–1820s.* Gainesville: University Press of Florida, 2011.

"Southern Boundary." See "Records Relating to the Southern Boundary of the United States," above.

Weber, David J., *The Spanish Frontier in North America.* New Haven, CT: Yale University Press, 1994.

Weeks, Charles A. *Paths to a Middle Ground: The Diplomacy of Natchez, Boukfouka, Nogales, and San Fernando de las Barrancas, 1791–1795.* Tuscaloosa: University of Alabama Press, 2005.

Whitaker, Arthur Preston. *The Mississippi Question: A Study in Trade, Politics and Diplomacy.* Gloucester, MA: Peter Smith, 1962.

——. *The Mississippi Question, 1795–1803.* New York: Appleton-Century, 1934.

Woodward, Ralph Lee Jr., ed. *Tribute to Don Bernardo de Galvez.* Translated from the Spanish by Ralph Lee Woodward Jr. Baton Rouge: Moran Industries, 1979.

Wilkinson, James. *Memoirs of My Own Times.* 3 vols. Philadelphia: Abraham Small, 1816.

Wright, J. Leitch, Jr. *William Augustus Bowles: Director General of the Creek Nation.* Athens: University of Georgia Press, 1967.

INDEX